Farmers' Cooperatives and Sustainable Food Systems in Europe

Farmers' cooperatives are very prevalent in the European Union, where they account for approximately half of agricultural trade and thus are key to articulating rural realities and in shaping the sustainability credentials of European food and farming. This book analyses to what extent farmers' cooperatives are working to benefit their members, are showing concern for their communities and are promoting cooperative economies. It offers a multilevel set of theoretical, disciplinary, methodological, empirical and social perspectives, using the UK and Spain as contrasting examples, and analyses whether agricultural cooperatives contribute to achieving sustainable food systems. The book presents empirical data from diverse and rich case studies, from large, international cooperatives, to small, multi-stakeholder initiatives. This provides an alternative viewpoint to that of economics, which tends to dominate the study of agricultural cooperatives. The author presents a new theoretical framework that provides a novel lens to study farmers' cooperatives as organisations deeply embedded in the power dynamics of the food system and agricultural policy that shape and constrain their potential to adopt cooperative and sustainable practices.

The book is a major addition to the study of agricultural cooperatives and their impact in the development of fairer and more sustainable food systems and it is one of the first detailed accounts of multi-stakeholder food and farming cooperatives in Europe. It is a valuable resource for all scholars working on cooperatives, as well as for students studying agricultural and food policy, environmental justice and rural sociology.

Raquel Ajates Gonzalez is a Researcher at the European Commission-funded GROW Citizens' Observatory led by the University of Dundee, UK. Before that, she worked as a Teaching Fellow at the Centre for Food Policy, City, University of London, UK, where she also completed her PhD.

Other books in the Earthscan Food and Agriculture Series

For further details please visit the series page on the Routledge website: http://www.routledge.com/books/series/ECEFA/

Farmers' Cooperatives and Sustainable Food Systems in Europe

Raquel Ajates Gonzalez

LONDON AND NEW YORK

First published 2018 by Routledge

2 Park Square, Milton Park, Abingdon, Oxon OX14 4RN
605 Third Avenue, New York, NY 10017

Routledge is an imprint of the Taylor & Francis Group, an informa business

First issued in paperback 2021

Publisher's Note

The publisher has gone to great lengths to ensure the quality of this reprint
but points out that some imperfections in the original copies may be apparent.

British Library Cataloguing-in-Publication Data
A catalogue record for this book is available from the British Library

Library of Congress Cataloging-in-Publication Data
Names: Gonzalez, Raquel Ajates, author.
Title: Farmers' cooperatives and sustainable food systems in Europe /
    Raquel Ajates Gonzalez. Other titles: Earthscan food and agriculture.
Description: New York, NY : Routledge, 2018. | Series: Earthscan food
    and agriculture | Includes bibliographical references and index.
Identifiers: LCCN 2017059173 | ISBN 9780815379249 (hardback) |
    ISBN 9781351216302 (ebook)
Subjects: LCSH: Agriculture, Cooperative—Europe.
Classification: LCC HD1491.E85 G66 2018 | DDC 334/.683094—dc23
LC record available at https://lccn.loc.gov/2017059173

ISBN: 978-0-8153-7924-9 (hbk)
ISBN: 978-0-367-51094-7 (pbk)

Typeset in Bembo
by Swales & Willis Ltd, Exeter, Devon, UK

Contents

Illustrations

Figures

Tables

Acknowledgements

I would like to thank all the people who agreed to be interviewed for this research for their generosity with their time and for being so kind as to share their thoughts and experience with me.

Thank you also to Martin Caraher and Tim Lang at the Centre for Food Policy, City, University of London, for their inspirational, lifelong dedication to improving food systems and for their fascinating reading lists. I would also like to thank Louise Hicks for encouraging me to pursue my studies in food policy. I am very grateful to City, University of London for the funding received to carry out the research project presented in this book.

Thank you to my parents for always prioritising my education. To my mum, for fostering in me and sharing a deep love of reading and books. To my dad, for teaching me that happiness is found in giving. Thanks to them and the inventor of the touch-typing method.

I would like to say thank you to my family and all my cherished friends for being so supportive, encouraging and understanding. Thank you to Maya Evans for being so brave, fun and inspiring. To Rebecca Wells, Kelly Parsons, Laura Favaro and other friends and colleagues at the Centre for Food Policy and City for being such incisive and caring critical thinkers.

I would like to say thank you and sorry to my body for spending so many hours in front of the computer.

Finally, the biggest thank you goes to Phil for his love, sense of justice, patience, ongoing support and trust. I am forever grateful.

I would like to dedicate this book to all the people who are giving up easy comforts in their lives to work towards fairer, regenerative food systems – thank you for keeping the seeds alive.

Abbreviations

AC	agricultural cooperative
ACCA	Agricultural Central Co-operative Association
ACF	Advocacy Coalition Framework
ACMA	Agricultural Co-operative Managers Association
AECA	Asociación Española de Cooperativas Agrarias
AOS	Agricultural Organisation Society
BSE	Bovine spongiform encephalopathy
CAP	Common Agricultural Policy
CCAHC	Central Council for Agricultural and Horticultural Cooperation
CEC	Common Land Cooperative
CMO	Common Market Organisation
CS	cooperative sustainability
CUMA	Cooperative for Common Use of Machinery
DEFRA	Department for Environment, Food and Rural Affairs
EC	European Commission
EIP	ecologically integrated paradigm
EFFP	English Farming and Food Partnerships
EU	European Union
FAC	Federation of Agricultural Co-operatives in Great Britain and Ireland
FCB	farmer-controlled business
FFB	Food from Britain
IAOS	Irish Agricultural Organisation Society
ICA	International Cooperative Alliance
ILO	International Labour Organization
IOF	investor-owned firm
ISEC	Instituto de Sociología y Estudios Campesinos (Institute of Sociology and Peasant Studies)
LSP	life-sciences paradigm
LWA	Land Workers' Alliance
MAGRAMA	Spanish Ministry for Agriculture, Food and Environment
MLG	multilevel governance

MSC	multi-stakeholder cooperative
MVP	Manchester Veg People
NEF	New Economics Foundation
NFU	National Farmers Union
NGC	new generation cooperative
OPA	Professional Agrarian Organisation
PG	Producer Group
PO	producer organisation
POF	privately owned firm
PP	productionist paradigm
SAOS	Scottish Agricultural Organisation Society
SAT	Sociedad Agraria de Transformación (agricultural processing society)
SD	sustainable development
SDC	Sustainable Development Commission
SFS	sustainable food system
UAOS	Ulster Agricultural Organisation Society
UCAE	Unión de Cooperativas Agrarias de España
UN	United Nations
UNDP	United Nations Development Programme
US	United States of America
USDA	United States Department of Agriculture
WAOS	Welsh Agricultural Organisation Society
WTO	World Trade Organisation
WWI	World War I
WWII	World War II

1 Introducing agricultural cooperatives in the context of a failing food system

Context, clashing definitions, principles and typologies

Despite earlier records of farming cooperatives, it was not until 1852 in Britain that the legal cooperative form entered the law for the first time in history (Zeuli and Cropp, 2004). During the nineteenth and twentieth centuries, diverse farmer cooperatives mushroomed across the European continent. By pooling resources together, farmers could maximise their purchasing power and acquire agricultural inputs of better standards. In post-war Europe, the creation of the Common Market marked a period of policy changes and a heated period of cooperative formation owing to the abolition of marketing boards and other protective governance measures that were deemed to interfere with EC competition law (Davey et al., 1976). Today, agricultural cooperatives (ACs) have a strong presence in the EU market, accounting for around half of all EU agricultural trade (Bijman et al., 2012). In 2011, the International Cooperative Alliance's list of the top 300 cooperatives revealed that many of the biggest European cooperatives operate in the agricultural supply or food and drink sectors (ICA, 2011).

Cooperatives, regardless of the sector they operate in, are expected to work for the benefit of their members, show concern for their communities (including sustainable development) and promote cooperative economies. This book analyses to what extent this is happening in the case of ACs. Evidence of how unsustainable and unequal farming in Europe is, despite such a strong AC presence, raises questions about the role and practices of these cooperatives. Despite their grassroots origins, concerns from civil society and a handful of scholars suggest there is an increase in top–down approaches and corporatisation trends in the sector. This book examines ACs in the context of the EU/Common Agricultural Policy (CAP) framework, examining how the sector has evolved since its beginnings and analysing trends and factors shaping their current development.

Going beyond the economic perspective that dominates the study of ACs, this research also focuses on emerging, innovative multi-stakeholder governance models. The strategies used to protect their alterity as well as the diverse understandings of food sustainability that different types of cooperative have and how they reproduce these through their practices are analysed. Given the insufficient explanatory potential of existing theories to accommodate a wide range of realities labelled as cooperatives in food and farming, a new theoretical framework was developed based on the findings of this research. The multilevel framework unravels the different dimensions that constitute cooperatives and their degree of alterity and commitment to sustainable food practices and the wider cooperative movement.

This book offers a critique of ACs beyond the dominant institutional economics lens, contributing to the debate on their de/repoliticisation. While acknowledging the important role ACs play in supporting farmers – many of those I spoke to during this research reported they could not survive without them – this book presents a solidary-critical approach (Favaro, 2017) through which to recognise and appreciate the often essential role ACs play in supporting individual farmer members, while still exploring difficult issues around co-optation that reveal a somehow uncomfortable, but potentially fruitful, analysis to help ACs move towards a fairer and more sustainable future.

Why study agricultural cooperatives? Relevance and timeliness

Why study ACs? Why is this topic relevant and worth researching? The history of food and farming cooperatives is intrinsically linked to the origins of the cooperative movement itself, although informal food provisioning practices based on cooperative relations long preceded the first formalised cooperative enterprises (Chloupkova et al., 2003; Sennett, 2012). The earliest records of cooperatives date back to 1750s France (Shaffer, 1999); soon after, other cooperatives emerged in Greece, Italy and Luxembourg (Shaffer, 1999; Naubauer, 2013). During the nineteenth and twentieth centuries, adulteration, quality and price of food and farming inputs were shared concerns that drove the creation of both consumer and agricultural cooperatives (Rhodes, 2012).

More recently, in post-war Europe, the introduction of the Common Market brought about dramatic policy changes and started a heated period of cooperative formation following the abolition of existing marketing boards and other protective governance measures deemed to interfere with EC competition law (Davey et al., 1976). Today, ACs are promoted around the world as organisational mechanisms to increase lobby power for farmers (Fairtrade Foundation, 2011). There are many types of AC, with aims from machinery-sharing to processing and marketing. A clear benefit of ACs is that they not only help farmers concentrate demand to buy cheaper inputs in bulk, but also allow them to negotiate better prices when selling produce to large buyers. Low prices are a real concern for farmers in the agricultural market, as the profitability of farming activities has been progressively declining for decades in favour of the processing and retail sectors. Additionally, by being members of ACs, small-scale farmers can benefit from access to training and technologies that they would otherwise be unable to afford (Giagnocavo, 2012).

Today, there are around 750,000 ACs across the world (Ortmann and King, 2007). Producer cooperatives are key to farming and account for a larger share of the cooperative economy than worker cooperatives (Wilson and MacLean, 2012). Agriculture is in fact the largest sector by annual cooperative turnover, with more than 39 per cent, or €347 billion, of the total annual cooperative turnover in Europe, followed by retail with nearly 30 per cent, or €264.38 billion (Cooperatives Europe, 2016).

There are tens of thousands of ACs in Europe, with Cogeca, the European body representing ACs, counting about 40,000 cooperatives on its books with a turnover of approximately €350 million (Cogeca, 2015). European ACs employ about 600,000 workers and have 9 million farmer members (Tortia et al., 2013). Large differences amongst member states exist, but, across the board, farmers' cooperatives are a key actor in the EU food system, supplying more than 50 per cent of agricultural products and more than 60 per cent of the collection, processing and marketing of agricultural products (Cogeca, 2010). The average market share of all ACs in the EU is 40 per cent, or higher if "hybrid cooperatives" with non-farmer investors are taken into the calculation (Bijman et al., 2012). It is not a black and white picture, but the tendency is that in the north there are bigger, more consolidated single-sector ACs, some of which are multinational cooperatives; in the south, more local, even multisectoral cooperatives operate, but still a handful of big ACs at the top account for most of the cooperative trade (Bijman et al., 2012). The overall trend seems to be for increasing concentration, with fewer but bigger cooperatives and a growing number of transnational ACs with members in more than one country (Bijman et al., 2012). The year 2012 was the United Nations (UN) International Year of Cooperatives, celebrated with the motto "cooperatives can feed the world". The question to ask was perhaps not whether ACs can feed the world, but at what price to consumers, the environment and small farmer members. This book investigates this question, trying to analyse why ACs are following a trend of growth and internationalisation and with what effects on their farmer members.

Their significance, not only in Europe but also across the world, is undeniable. However, despite their strong presence, the European farming problem is obvious at different levels:

- There is an increasingly high market concentration at producer, processor and retail level that creates power imbalances in the food system and asymmetrical bargaining powers (e.g. UK farmers only receive around 7 per cent of what consumers spend on food (Pretty, 2001), and around two-thirds of all food is sold by only four supermarkets (Morgan et al., 2015). In Europe, the top five retailers' market share at national level (not necessarily the same five in each member state) exceeds 60 per cent in 13 member states (European Commission, 2014a).
- Concentration is also present in subsidy payments: 80 per cent of CAP subsidies go to 20 per cent of farmers, mainly for large, monoculture exploitations (BBC, 2013).

- Agricultural land availability and affordability are rapidly decreasing: half of all farmland in the EU is now concentrated in the 3 per cent of large farms bigger than 100 hectares in size (European Union, 2012). At the same time, landownership is also concentrated (e.g. 70 per cent of UK land is owned by less than 1 per cent of the population (Home, 2009).

- Evidence of the environmental impacts of farming is piling up. Although statistics on the environmental impacts of the food system vary significantly depending on the source and the method of calculation (e.g. land cleared to cultivate feed, transport, etc.), they account for between 18 and 30 per cent of all greenhouse gas (GHG) emissions related to human activity (Garnett, 2014). Another sobering statistic is that around 20 per cent of global GHG emissions are produced by the animal husbandry sector alone (Steinfeld et al., 2006). These statistics reflect the impact of food production methods that are unsustainable both for the environment and for farmers (UNEP, 2009).

- Concerns over agricultural labour shortages and an ageing workforce are growing every year (European Union, 2011). One-third of European farmers are over 65 years old, and only 7 per cent are under 35 (European Commission, 2013a).

- There is a generalised disconnection of farming from consumers and the environment. Long supply chains mean consumers are very removed from their food and the people who produce it (Kneafsey et al., 2008). Inequalities in extent and quality of consumption have fuelled the spread of diet-related diseases and food banks (Gentilini, 2013).

This picture of unsustainable and unequal farming in Europe despite such a strong AC presence raises questions about the role and practices of these cooperatives. If ACs are supposed to represent members' interests, why are the above challenges facing farmers becoming so acute? Cooperatives are not a legacy of the past, but more and more, especially since the beginning of the 2007–08 financial crisis, they seem to be a popular model for new enterprises, whether under conservative or socialist administrations (COCETA, 2012; CECOP, 2013; Co-operatives UK, 2013b). The year 2012 was the UN International Year of Cooperatives, from which the Blueprint for a Cooperative Decade emerged, setting up the vision for the movement until 2020. This international focus on cooperatives, coupled with the increasing interest and concerns about sustainability in food systems, makes the topic of this research invaluably timely.

Additionally, in terms of challenging taken-for-granted assumptions (Tracy, 2010), this research offers an insight into the world of ACs with findings that clash with well-established ideas or imaginaries about cooperatives and cooperative members. Evidence shows people have a positive view of cooperatives as more ethical businesses (Cooperativas Agro-alimentarias, 2013), many reporting a willingness to buy more or pay more for cooperatively produced items, as they believe there is an added ethical value in the way cooperatives operate (Cooperativas Agro-alimentarias, 2013). As this book will present, this imaginary can be no more than a halo effect in the dominant AC sector.

Gaining understanding of the trends and dynamics within the AC sector is instructive; however, the utility of such an analysis is limited if it cannot be understood within the broader context of global food and farming systems in which the sector operates, as well as the pressing issue of food sustainability and the wider cooperative movement from which ACs emerged. This contextualisation has been addressed through a detailed overview of the history of the movement, a review of the literature on sustainable diets and current data on food and farming and ACs in Europe and a discussion of ACs' links with other social movements in Chapters 8–9.

The cooperative movement has strong, diverse roots, and ways of cooperating in food and farming are evolving in diverse ways. The fieldwork revealed a flourishing richness of governance forms and cooperative sizes in the AC sector. This forms the crux of the research problem. The analysis of the nexus between this richness and how these practices have an effect on bringing closer or distancing the achievement of fairer and more sustainable food systems is the focus of this book.

Structure of the book

This book documents the historical development and current picture of the AC sector in Europe, using Spain and the UK[1] as case studies, and how ACs are being shaped and misshaped by and within the European farming policy context and the architecture of global food governance. The research presented is interested in the social and economic conflicts and contradictions that ACs are encountering and how this is having an effect on their members, social justice in the food system and the environment. The book covers a wide area of empirical and theoretical ground. Theoretically, it provides a multidisciplinary review of agricultural cooperatives in economics, history, psychobiology and food policy literature, as well as analysing data from European and national governments' reports.

The book starts by unpacking the term cooperative in this first chapter, its more common definitions in the context of agriculture, the importance of the cooperative principles, and the different typologies used to categorise ACs. To put the research into a historical context, the evolution of ACs in Europe is presented in Chapter 2, with consideration of the social, political and policy factors that at different times in history either fostered or hindered the formation of ACs. The current picture of the scale of the AC sector in the twenty-first century and its role in European farming is then reviewed, with a particular focus on fruit and vegetable cooperatives.[2] Attention is then shifted to the literature, with a critical, comprehensive review of the different disciplines that have entertained cooperatives, and presentation of the argument that the way they have been studied has created a path-dependent approach to measuring success in ACs. The second part of Chapter 3 reflects on the challenges and ongoing efforts to define sustainable food systems and the dis/connections with cooperative organisations. A description of the theoretical influences and how the selection of theory and methods matters when studying cooperativism and food sustainability is provided in Chapter 4.

Given that the findings revealed a huge diversity of ACs, with a varying degree of integration in industrial globalised food systems and others more embedded in alternative food networks (AFNs), a series of case studies are presented to illustrate the differences. The book covers two country cases: UK and Spain, in Chapters 6 and 7. These countries have contrasting cooperative and agricultural development histories, but some recent policy similarities convert them into two gripping cases that unveil how the politics of the past affect present cooperative arrangements. Chapter 5 presents and discusses experts' views on current and future trends for ACs at the European level. Within each case country, a variety of ACs are examined, starting with the history and current policy context for ACs in Chapter 6, followed by the study of two commercially successful cases in Chapter 7. New emerging models of multi-stakeholder cooperation are discussed in Chapters 8 and 9.

All the case studies cover the social, economic and sustainability dimensions of the ACs presented, always framed in the context of European farming policy. Some of the case studies that will be presented are legally incorporated cooperatives that bring together growers, buyers and eaters under the same legal entity. As opposed to conventional agricultural cooperatives with a single type of member (e.g. farmers), the multi-stakeholder cooperatives (MSCs) presented in Chapters 8 and 9 bring different groups of stakeholders as distinct membership groups under one shared cooperative enterprise to meet their diverse food-related needs. For this reason, a more cooperative-specific theoretical framework suitable to analyse the richness of these innovative governance models was required. The Open Cooperative model from the P2P Foundation was selected for this purpose, based also on its explanatory potential to take into account these MSCs' links with the wider cooperative movement, other social movements and their efforts to create more sustainable food systems.

The concept of "third space" is then used in Chapter 9 to pay attention to the strategies devised by a new generation of cooperatives in order to protect their alterity and their transformative potential. These are theorised in relation to the creation of physical and metaphorical third spaces in which new identities and ways of cooperating in food production and provision are negotiated and adopted. As this book argues, third space, a concept borrowed from colonial studies, can be applied as a powerful construct that adds fluidity to the alternative/conventional binary that currently has a stronghold in food studies. Additionally, it opens up theoretical room to manoeuvre between food production and consumption issues.

Given the limitations of the economic discipline to study the socio-political and sustainability dimensions of ACs, the findings of this research are incorporated into a new theoretical framework presented in Chapter 10. The framework developed holds sufficient explanatory potential to accommodate and examine a wide variety of initiatives labelled as ACs. Considering this variety of cooperative forms and cooperative practices was not an empty theoretical exercise but an attempt to capture the richness revealed by the fieldwork. The framework accounts for and reflects a process of quantification, fragmentation and appropriation of the agricultural cooperative sector by the dominant food regime that explains ACs' loss of transformational potential. The book concludes with Chapter 11 summarising the outcomes of the study, reflecting on the research process, identifying areas for future research and considering implications for farming policies.

Methods

This book presents the result of a research project that aimed to study the diversity of producers' cooperatives operating in food and farming in Europe and their relationship with food sustainability. In order to capture as much as possible of the complexity of the interconnections between both topics, cooperation and sustainability, a multilevel qualitative approach was followed:

1   *Macro*: national and European level (Phase 1).

The top, macro level focused on national and EU policies (also in the context of transnational WTO agreements) relevant to cooperatives and food policy. This level related to the role public policy has in shaping ACs' practices and their impact on ecological public health and social and environmental justice. It identified and reviewed the policy and contextual factors affecting ACs. Macro data included AC statistics from government, EU and industry reports, as well as academic literature and public policy documents.

2   *Meso*: organisational case study (Phase 2).

The meso level focused on the cooperative organisation in order to:

- identify a spectrum of ACs according to size, different levels of integration of non-farmer actors (e.g. consumers and buyers, as in the case of emerging multi-stakeholder models) and level of embeddedness in the industrial food system;
- select case studies according to their location on the above spectrum, taking into account it is not linear, and that many hybrid ACs exist;
- reflect the diversity of the sector. For theoretical purposes, diverse examples of ACs were chosen. The literature reveals a trend towards consolidation and specialisation in an increasingly concentrated cooperative sector. At the other end of the spectrum, a growing number of smaller ACs with a more social and environmental focus started to appear during the 2007–08 financial crisis, fuelled by the desire to make business more ethical and to return to closer-to-nature ways of living. The case studies represent this

variety characteristic of the AC sector: from large ACs with international operations (discussed in Chapters 5 and 7) to those trading locally only but being politically active globally (discussed in Chapters 8 and 9); variety was also reflected in terms of size and governance structures.

3   *Micro*: Individual accounts of cooperative members' experiences (Phase 3).

This final level of analysis aimed to:

- collect and analyse cooperative members' points of view about the role of ACs and their reasons for joining;
- investigate the subjective views and experiences of members and policy-makers regarding differences amongst ACs;
- explore how members understand food sustainability.

In total, 41 interviews were conducted including both with ACs' members and other relevant actors. Although all were included in the analysis, not all are specifically referenced or quoted in the text. The interviewees fell into one or more of the following categories:

- 1 industry;
- 1.1 large cooperatives;
- 1.2 multi-stakeholder cooperatives;
- 1.3 industry representative bodies/unions;
- 2 government;
- 3 academics;
- 4 civil society.

Interviews took place in 2014–15, and the research covers an interesting period for the cooperative sector (2012–16), as new cooperative experiments had emerged just before, encouraged by a motivation, and in some cases an urgent need, to try new models of food provisioning due to the effects of the financial crisis that started in 2007.

Definitions, principles and typologies of agricultural cooperatives

There are historical records of cooperative enterprises dating back to the 1750s (Shaffer, 1999; Naubauer, 2013). The renowned Rochdale Society of Equitable Pioneers' cooperative grocery store, set up near Manchester (UK) in 1844, is considered the first successful cooperative (Sanchez and Roelants, 2011). This early consumer cooperative was the first one to pay dividend payments to members and to establish a well-developed set of statutes and principles (Fairbairn, 1994). Since Rochdale, cooperatives have evolved in many ways and shapes, have flourished and failed in many sectors, and have reflected the needs and principles of their societies at different times in history. This research focuses on agricultural cooperativism that has in itself a myriad of different forms. A few statistics will give the reader a rough picture of the global importance of ACs: roughly half of all agricultural products are traded through cooperatives (Bijman et al., 2012); approximately half of all British farmers and 85 per cent of Spanish farmers are members of at least one AC (two country cases that will be discussed in more detail in the book); and across the world, three-quarters of Fairtrade goods are produced by ACs (Cooperative Heritage Trust, 2012). Next, this chapter will provide an introduction to cooperative definitions and principles and how these affect cooperative practices in real life, as well as analysing the more common typologies found in the literature.

What is a cooperative?

The internationally accepted definition of cooperative comes from the International Cooperative Alliance (ICA):

> A cooperative is an autonomous association of persons united voluntarily to meet their common economic, social, and cultural needs and aspirations through a jointly-owned and democratically-controlled enterprise.
>
> (ICA, 1995)

Key words to highlight from the ICA's definition are: autonomous, needs and aspirations, as well as the ICA's focus on democracy and enterprise, aspects of the cooperative model that clash with the capitalist economies cooperatives exist in, as basically, a cooperative is "people-centred" rather than "capital-centred". The latter is the defining feature of investor-owned firms (IOFs) or private companies (Birchall, 2010).

The cooperative principles are guidelines by which cooperatives put their values into practice and are what differentiate them from IOFs. The ICA proposes seven principles based on those of the original Rochdale Pioneers. The influence of the Rochdale Cooperative in the global cooperative movement is still strong today. When discussing the ICA cooperative principles, it is pretty much obligatory to mention the Rochdale cooperators in order to understand the roots of the current ICA principles. These were the original principles agreed by the Rochdale cooperators in 1844:

- Voting is by members on a democratic basis (one member, one vote).
- Membership is open.
- Equity is provided by patrons.
- Equity ownership as share of individual patrons is limited.
- Net income is distributed to patrons as patronage refunds on a cost basis.
- Dividend on equity capital is limited.
- Goods and services are exchanged at market prices.

The above principles were written in 1844, when the cooperative was formed, and were later accepted by the ICA in 1937, remaining unchanged until 1966. The ICA only modified the principles on one other occasion, in 1995, in order to adapt them to the present context, with the final and current version going beyond the business sphere and taking into account the social character and social responsibility of cooperatives (ICA, 1995). Since 1995, the seven internationally recognised cooperative principles have been:

1   voluntary and open membership;
2   democratic member control;
3   member economic participation;
4   autonomy and independence;
5   education, training and information for both members and the general public;
6   cooperation among cooperatives to strengthen the cooperative movement;
7   concern for the sustainable development of their communities.

Cooperatives in all sectors are expected to adhere to the seven internationally recognised ICA cooperative principles. These principles, in theory, are there to inform cooperatives' practices and to differentiate them from privately owned business (ICA, 2015). However, during the 1980s and 1990s, before the publication of the latest version of the ICA principles, global markets, including for food, started to become increasingly open, a trend that culminated in the incorporation of agriculture into the WTO in 1994–5. At that time, arguments for a simplification of the principles were put forward (Birchall, 2005) in order to "allow" ACs to attract large farmers to become more competitive. Critics suggested that the ICA principles were too restrictive, and claimed that they were only guidelines for members to choose from and decide which principles they wished to adhere to (Zeuli, 2004). In the US, the rapid rate of conversions of ACs to private companies led to some efforts to emphasise the differences between ACs and private companies, with the aim of reinforcing the cooperative identity and highlighting the cooperative advantage and the value of being member-focused (Dunn, 1988; Barton, 1989; Torgerson et al., 1997). Nevertheless, in 1987, the United States Department of Agriculture (USDA, 2002) adopted the following three principles for ACs (Ortmann and King, 2007):

1   The User-Owner Principle: Those who own and finance the cooperative are those who use the cooperative.
2   The User-Control Principle: Those who control the cooperative are those who use the cooperative.
3   The User-Benefits Principle: The cooperative's sole purpose is to provide and distribute benefits to its users on the basis of their use.

These principles are relevant to the European case as they were imported and adopted in the EU, as it will be discussed in more detail in Chapters 2 and 3.

Although the user principles excluded the education, community and wider cooperative movement principles of ICA, it still championed them as "practices" that could facilitate a deeper application of the three principles (Dunn, 1988). However, 10 years later, in 1997, the Rural Business Cooperative Service of the USDA published its Research Report 151 titled, "Strengthening ethics within agricultural cooperatives". The need for this report was identified after a series of meetings with industry at which ongoing concerns about the weakening ethics in farmer cooperatives were raised (USDA, 1997). This USDA research report reinforced the skimmed version of the cooperative principles, stating that:

> Ethical decisions in cooperatives, then, should reflect recognition and commitment to the user-control, user-benefit, and user-owner principles. Decisions that violate these three central tenets should be viewed as unethical.
>
> (USDA, 1997:11)

In 2014, the USDA published a new research report (number 231) comparing the cooperative principles of the USDA and the ICA and coming up with some interesting reflections (USDA, 2014). In this review, the USDA author, Bruce J. Reynolds, an economist, discusses how the three USDA principles were a "reduced form approach" developed through the "lens of economics" and prepared by economists to exclude values from the definition and identification of cooperatives (USDA, 2014:3). The resulting paradox is that removing values from cooperative practices in order to emphasise an economic edge dilutes, as Reynolds pointed out, the ethical and social essence of the cooperative identity:

> Economists avoid asserting social values, yet this cautiousness when applied to principles may exclude attributes that contribute to further differentiating cooperatives from other forms of organization.
>
> (USDA, 2014:3)

Evidence of the impact that the dilution of cooperative principles can have on members and food sustainability is discussed throughout the book, and especially in Chapters 5, 6 and 10.

Types of agricultural cooperative

Cooperatives exist in many different sectors of the economy. From housing to banking, they have appeared over the decades to cover the very diverse needs of changing societies. As it is beyond the reach of this chapter to introduce every type of cooperative, Figure 1.1 has been used to summarise cooperatives based on the role of their members. In the case of food retail cooperatives, consumer members can act as voluntary employees; in some cases, this time commitment is a requirement of the cooperative, for example, in the famous Park

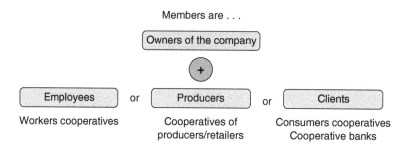

Figure 1.1 Typology of cooperatives across sectors based on members' role.
Source: Author

Slope Cooperative in New York (Ronco, 1974; Jochnowitz, 2001); in others, members exchange work for discounts on transactions with their cooperatives or simply use it as a way of celebrating and learning about food (Actyva, 2015).

There is a wide range of entities in food and farming labelled as cooperatives, from producers working together in more informal, not legally incorporated associations (Vara Sanchez, 2008), emerging agricultural workers' cooperatives, multi-stakeholder models in which consumers also become members (Gray, 2014b; Ajates Gonzalez, 2017a) and community-supported agriculture schemes, to small village cooperatives that refuse to merge. A more recent model that is growing strongly is that of Producer Organisations (POs) that will be discussed in more detail later (Bijman et al., 2012).

With regards to ACs, many typologies, mainly coming from the economics discipline have been attempted, most of them based on their ownership structures and financial models. Looking at Figure 1.1, it can be said that sometimes the boundaries for agri-cooperatives are not clear, as worker cooperatives cultivating shared plots of land exist; recently, there has also been a trend of new cooperatives linking producers with consumers in an attempt to set up shorter supply chains. These relatively new models will be analysed in detail in this book, and, therefore, when ACs are referred to throughout the text, the term is used in an encompassing manner to include both conventional ACs (supply, machinery, processing and marketing) and agricultural workers' cooperatives and multi-stakeholder cooperatives that bring together different memberships into the same cooperative organisation (growers, workers and consumers and/or buyers).

Table 1.1 summarises the different categorising criteria often used in the literature to classify ACs. Existing typologies often avoid categorising ACs by type of product/service or sector in which cooperatives operate, opting instead for a categorisation based on investment or governance structures. However, the last comprehensive report on European marketing cooperatives from the European Commission, published in November 2012, highlighted that the character of the product, such as perishable products that benefit from processing (e.g. milk) or a swift sale (e.g. fresh vegetables and fruits), influences the investment policies of its associated cooperatives: processing infrastructure for the former example and logistics for the latter (Bijman et al., 2012). This report also revealed that the degree of specialisation of cooperatives is on average higher in north-west Europe than in southern Europe, a difference that will be explored further in relation to the country studies of Spain and the UK (Bijman et al., 2012). The criteria presented in Table 1.1 were compiled by the author to highlight the many existing ways of categorising ACs. These typologies are not exclusive and can overlap.

Table 1.1 Common categorising criteria found in the literature used for the classification of agricultural cooperatives.

Categorising criteria	Associated typology
Cooperative principles	Rochdale, traditional, proportional and contemporary principles (Barton, 1989).
Ownership and financial structure	Several variations of the dichotomy between traditional (production-oriented) versus new generation cooperatives (market-oriented; Van Bekkum, 2001; Kyriakopoulos et al., 2004; Cook and Plunkett, 2006).
Types of member	Primary (members are individual farmers) and secondary (members are primary cooperatives). Secondary cooperatives normally have both economic and lobbying functions (Bijman et al., 2012).
Legal form	Association, partnership, cooperative, limited liability company, etc. (e.g. POs and SAT benefit from CAP cooperative subsidies but can have any legal form).
Ideology	More common in historical literature. From communist to libertarian, nationalistic and capitalist cooperatives, as well as those with underlying religious beliefs (e.g. Longo Maï and Israeli kibbutzim).
Function	• Production • Inputs supply • Processing • Marketing • Insurance
Position in the food chain and degree of vertical integration	• Primary processing • Secondary processing • Retailing (direct sales)
Degree of internationalisation (both for business activities and geographical scope of membership)	• Local • Regional • Interregional • National • Transnational

(continued)

Table 1.1 (continued)

Categorising criteria	Associated typology
Strategy	Traditional vs. entrepreneurial (Kyriakopoulos et al., 2004)
	Cost leadership, differentiation and organisational structure (Van Bekkum, 2001)
Degree of specialisation	*"Manymixed"* (non-specialised) vs. specialised in one product (Spear et al., 2012)
Life cycle	From formation to dissolution or restructuring/demutualisation (Cook, 1995).
	A later cycle identified the following stages (Cook and Plunkett, 2006):
	1. Genesis
	2. Growth
	3. Emergence of internal conflicts,
	4. Recognition and analysis
	5. Option choice

Source: Author

Being informed by the cooperative principles and being co-owned by members, cooperatives are complex organisations, as their main objective distances the organisation from the conventional dominant aim of firms: profit-making. The multiple stakeholders of cooperatives (members, other cooperatives and communities), as well as their praxis (democratic governance), create investment constraints. From an economics perspective, many authors have analysed these challenges, summarising them into four commonly cited cooperative problems (Nilsson, 1999; Iliopoulos, 2005):

The first three refer to investment constraints:

1 **Horizon problem**: this arises when members prefer higher dividends now rather than investing in the cooperative for the future, basically neglecting the long-term competitiveness of their organisations.
2 **Free rider problem**: this problem is common when some members do not contribute the same as others but still enjoy the same benefits. It can also occur when new members do not increase competitiveness significantly but enjoy the work and efforts of earlier members, leaving both groups without a logical reason to make new investments or improvements.
3 **Portfolio problem**: different members will bring different products of different quality to the cooperative, creating problems around how to best market them but also for the farmers, especially older ones (the bulk of current farmers), who might prefer to invest part of their portfolio away from the cooperative in order to secure better returns and redeemable savings.

The last, most cited problem does not refer to an investment barrier, but to the transactional costs of the cooperative model's democratic approach:

4 **Collective decision-making costs:** this problem arises from democratic processes, but also from lack of management training and the difficulties of monitoring a manager who often is not a member of the AC. Additionally, the "influence cost constraint" refers to the negative effect of unbalanced influences of members on decision-making (Iliopulos, 2005).

As the reader will notice, the above problems focus on an organisational dimension that does not consider ACs in the wider policy context or cooperative movement, a gap this book will address. Despite these four common problems, for a long time cooperatives have been championed as a global solution for farmers across the world; 2012 saw an increase in the promotion of ACs, as it was the UN International Year for Cooperatives. In food and farming, with the increasing growth and concentration of first the processing sector and later the retail sector, along with the opening of global agricultural markets, reasons for encouraging the development of ACs include (Bijman et al., 2010):

- overcoming market failures;
- accessing markets (both national and international);
- gaining economies of scale for inputs, concentration of production and processing;
- strengthening bargaining power with suppliers and customers;
- reducing transaction costs and making markets more transparent;
- pooling resources to fund innovation.

Nowadays, cooperatives cite counterbalancing power with retailers as a stronger reason for cooperation than economies of scale (Bijman et al., 2010).

Summary

ACs are promoted around the world as organisational mechanisms to increase lobby power for farmers, access to machinery and reduced prices on inputs and other farm costs. There are many types of AC, with aims from machinery-sharing to processing and marketing. Despite the strong presence of ACs in Europe, the social, environmental and economic challenges facing EU farmers and consumers are huge. This chapter posed the following question: Why are cooperatives unable or unwilling to address these challenges?

A summary of definitions and typologies of agricultural cooperatives in the academic literature, primarily from the economics discipline, has been provided. As the numerous existing categorising criteria show, cooperatives can be studied and analysed in many ways. This assortment of "lenses" through which to look at cooperatives has affected the way their performance has been

measured, as will be discussed in Chapters 10 and 11. The cooperative principles introduce an added layer of complexity to the *raison d'être* of ACs, pushing
them beyond the traditional, monolithic profit-making aim of private firms.
This chapter has discussed in detail the evolution of the ICA cooperative principles and their dilution in the AC context and government reports. The next
chapter discusses the history of ACs in Europe, covering their development
from the origins of the first cooperatives to the present day.

Notes

1 The rationale and method for selecting these two countries are explained in Chapter 4
  and summarised in a table included in Appendix I.
2 One of the theoretical assumptions of this book is the need for integrated food policy
  framed by an ecological public health perspective that emphasises the need for a food
  system that is beneficial to both health and the environment and that is founded on
  plant-based diets (Lang et al., 2009). This approach is discussed in detail in Chapters
  3-4.

2   Past and present

The evolution of agricultural cooperatives in Europe from the nineteenth to the twenty-first century

The cooperative movement began in the eighteenth century with the realisation that the power of organised cooperation had the potential to transform society and shake structural conditions causing inequalities (Shaffer, 1999). Many cooperatives were created (and still are) as a response to difficult economic situations and complex social problems affecting the lives of those involved (Fairbairn, 2004). As Fairbairn has pointed out in relation to the Rochdale cooperators, most people at the time had "no vote, no democratically elected government to represent them, no interventionist state to protect them" (Fairbairn, 1994:2); in this dire context, the Rochdale pioneers found an answer in the practice of self-help and mutual help that cooperation offered them.

At the beginning of the eighteenth century, before cooperatives were born, two types of organisation offered European citizens a channel to participate in their free societies. First, "voluntary associations" were formed by groups of people who shared the same public or community aim (Merrett and Walzer, 2004). These associations were democratic organisations whose primary objective was to have an active participation in society. Second, "business enterprises" were a second type of organisation, also of a voluntary character, used by members to participate in business activities within the free economy. As a result of democratic, libertarian and equalitarian thinking and the combination of these two existing types of organisation, the cooperative model started to develop (Fairbairn, 2006). Cheese makers' cooperatives in Franche-Comté (France) were the first documented producer cooperatives, dating back to the 1750s (Shaffer, 1999). The establishment of other cooperatives in different trades (farming, fire insurance, watchmakers, dairy, etc.) followed rapidly in Europe and the US (Shaffer, 1999).

Two French authors were responsible for some of the earliest writings on agricultural cooperativism: Saint–Simon worked on various theories of "associations", and Charles Fourier published a Treatise on Domestic Agricultural Association in 1822 (the first one of its kind). A few decades later, thanks to the work of Friedrich W. Raiffeisen, rural credit cooperatives were founded in Germany, but soon were replicated in other countries, also very successfully.

In 1864, Raiffeisen developed the first cooperative lending bank to help small farmers. Later, he founded a regional cooperative bank and a national cooperative bank and, in 1877, he unified the entire system, forming the first credit union (ICA, 2012).

In the UK, adulteration was a common concern that fuelled the growth of consumer cooperatives (Rhodes, 2012). In 1884 France, the *syndicats agricoles* were originally formed for buying manure and seeds, also because of farmers' growing concerns about the falsification and adulteration of these inputs. Soon after, these early cooperatives also introduced the sale of produce and improvements for the land (Kropotkin, 1902). Prince Kropotkin's famous book *Mutual Aid* was highly influential at the time, arguing that cooperation, not only competition, was very common in nature among many species (Kropotkin, 1902). Both Darwin and Kropotkin agreed that cooperation had helped higher evolution; however, Kropotkin regretted how Darwin had emphasised the competitive over the cooperative behaviours that occur in the natural world. For Kropotkin, both were laws of nature. Kropotkin's work is discussed again in more detail in Chapter 3.

The year 1844 was key for the movement. In this year, the Rochdale Pioneers in Manchester started a set of collective actions that shaped the first formalised cooperative in the world; their legacy is still highly influential today (Cook and Plunkett, 2006). A second significant event in that same year marked one of the earliest organised mobilisations for participatory social change: in Rodding, Denmark, the first folk high school for peasants opened, with the aim of educating the lower classes to become active citizens and change society from the bottom (Zeuli and Cropp, 2004).

Throughout the twentieth century, cooperatives became closely associated with syndicalism and communism. However, Karl Marx, one of the early communist theorists, actually believed that cooperatives and trade unions did not have "in fact any great social future, any part to play in the renovation of society", being only "'embryos' of the great workers' companies" that Marx hoped one day would replace capitalism (Coates and Benn, 1976:21); however, Marx acknowledged that cooperatives could pave the way for the transition to a new society (Marx, 1866).

In Europe, after the Great Depression at the end of the nineteenth century, governments adopted an increasingly interventionist approach to agriculture during the twentieth century in order to both improve yields and introduce technical innovations (Koning, 1994). This would affect the development of an independent agricultural movement over the last century, and, as a result of socio-economic and political influences, the label "cooperative" evolved in varied ways and with very different connotations across different European countries. As an example, Finnish agricultural cooperativism developed as a movement for independence in Russian-occupied Finland, and the model still has nationalistic connotations in this country today (Bijman et al., 2012). Other European nations have seen periods where associationism was restricted, as bottom–up cooperatives were considered to be a dangerous nest for dissident ideas – for example, during Salazar's regime in Portugal and Franco's dictatorship in Spain (Cervantes and Fernandes, 2008). In other European regions, cooperatives went through periods of being imposed from the top down, no longer based on self-organisation principles but used as a socialist planning tool instead, as happened during communist regimes in Eastern Europe (Bijman et al., 2012). At the European level, the representative bodies of cooperatives – both for farmers (Cogeca) and consumers (Euro Coop) – have always been given importance in formal EU relations since the beginning of the Common Market, emphasising the collaborative model of the union. This influence has been reflected over time in their inclusion in committees (O'Connell, 1980; Young, 1995) and also in influencing CAP negotiations (Egdell and Thomson, 1999).

Agricultural cooperation in the twenty-first century

First, in this section, a short introduction to agriculture in the EU and the CAP is provided. Then, the rise of POs and their effect on ACs is discussed.

In December 2012, a detailed review of ACs in the European Union was published. The European Commission commissioned Wageningen University to lead in this project, named "Support for Farmers' Cooperatives". As well as a final report, the project generated 78 background reports with the aim to support "the policy making process" and the Commission's efforts to encourage the creation of agricultural producer organisations in the EU (Wageningen, 2013). Given their depth and spread, data from these reports will be drawn on in discussions of the current picture of cooperativism in EU farming; the main project publication will be referenced as "Bijman et al., 2012", as those were the authors who wrote the synthesis report. The country reports will be referenced separately.

Agriculture in the European Union: the Common Agricultural Policy (CAP) and agricultural cooperatives (ACs) in the EU

Agriculture, fishing and forestry account for 1.8 per cent of European GDP (European Commission, 2012b) and 5 per cent of employment (European Commission, 2013a). Agricultural policy is framed by the CAP. Following food shortages during and after World War II (WWII), the CAP was created to ensure European citizens could enjoy affordable food and farmers could make a living. When the Treaty of Rome was signed on 25 March 1957, it already contained the most important framework provisions for the adoption of the CAP. At the time, policy approaches based on and promoting agricultural exceptionalism (the idea that food production is different from other economic sectors and as such deserves government support) shaped the development of the state-assisted paradigm in the US and Europe (Skogstad, 1998).

The Treaty did not stipulate any specifics with regard to the relationship between the European Community authorities and the representatives of the agricultural sector, but the European Commission soon expressed its desire for

Table 2.1 Early thinkers that inspired the cooperative movement

Thinker	Theoretical perspective	Period	Key ideas relevant to the emergence of cooperative movement
J. J. Rousseau, J. S. Mill	Participatory democracy	18th c. (JJR) 19th c. (JSM)	Citizen participation is essential for democracy, not just in voting, but in other spheres of society.
Charles Fourier	Utopian socialism (Fourierism)	Late 18th c. to early 19th c.	A new, fairer society based on *phalanges* (communal associations of agricultural producers).
William King	Utopian socialism, economic philosophy	Early 19th c.	Focus on consumer cooperatives to change habits, as shopping is an everyday activity. Concept of 'habit' as a barrier but also as a way of promoting long-term cooperation.
Charles Dunoyer	Economic philosophy	Early 19th c.	In 1830, French liberal economist Dunoyer published a *Treatise on social economy* that advocated a moral approach to economics, being the first time the term 'social economy' appeared in economics literature.
Robert Owen	Utopian socialism (Owenism)	19th c.	Use of cooperation to create a New Moral World. Capital does not create value, labour does.
Karl Marx, Friedrich Engel	Communism	19th c.	Cooperatives only useful as a transition to communism.
George Holyoake	Secularist Owenism, Chartism	19th c.	Cooperation as prevention of class war and stabilisation of society.
Pyotr Kropotkin	Anarchism, evolutionary theories	Late 19th c. to early 20th c.	Mutual aid in nature: not just competition, but also cooperation.
Robert Blatchford	Social imperialism, economic nationalism	Late 19th c. to early 20th c.	Britain for the British. Industrial cooperation for the common good.
George D. H. Cole	Libertarian socialism, participatory democracy	Early 20th c.	Cooperatives are a key area for practical anarchism. Alternative to oppressive power relations.

Source: Author

close cooperation with sector representatives. Aware of the importance of the Community to their sector, farmers' organisations soon organised and, on 6 September 1958, the first European representative farmers' organisation, Copa (Committee of Professional Agricultural Organisations), was created (Cogeca, 2015). One year later, on 24 September 1959, Cogeca (General Committee for Agricultural Cooperation in the European Union) was founded as the European umbrella organisation for ACs, also including fisheries' cooperatives. When Cogeca was formed, it had six members. At the time of writing, it had 30 full member organisations and one affiliated member from across the 28 EU member states. In addition to that, Cogeca has another 36 partner members (including ten coming from non-EU member states; Cogeca, 2015).

First introduced in 1962, the CAP is one of the oldest policies of the European Union and was the EU's first and, for many years, only fully integrated policy (European Commission, 2012a, 2013b). The first objectives of the CAP, still current today, were increased food security, market stabilisation and product support. With the 1992 reform, the CAP also tried to achieve income and budget stabilisation and increased competitiveness and started to look at environmental objectives. From Agenda 2000 onwards, there has been a clear emphasis on sustainability as a policy concern above all other objectives (European Commission, 2013b). The evolution of the CAP has witnessed a clear process of 'policy stretching', by which farming policy is expected to meet many objectives, including economic objectives, land management, environmental concerns, rural development and so on (Feindt and Flynn, 2009).

Every CAP reform cycle generates heated negotiations and public debate about the extent and role of farming subsidies. Nevertheless, agriculture only receives less than 1 per cent of public expenditure in the EU (Copa-Cogeca, 2012). Large differences exist among member states, but, across the board, ACs are key actors in the EU food system, being responsible for more than 60 per cent of the collection, processing and marketing of agricultural products (Cogeca, 2010). The average market share of all ACs in the EU is 40 per cent or higher, if including the trade of 'hybrid cooperatives' (which allow non-farmer investors) into the calculation (Bijman et al., 2012).

The extensive 2012 report commissioned by the European Commission into agricultural cooperatives in the EU sketched a few outlines of the main differences across member states: in northern Europe, ACs have become increasingly corporate; in the south, the sector is, on average, still quite atomised and localised; in the east, farmers are still reluctant to trust the cooperative model after the socialist experiments in their countries' recent history (Bijman et al., 2012). The largest European ACs are Dutch (Coforta/The Greenery), Italian (Conserve Italia), German (Landgard) and Spanish (Anecoop; Bijman et al., 2012:51). It is not black and white, but the tendency is that, in the north, there are bigger, more consolidated, single-sector ACs, even multinational cooperatives; in the south, there are more local, even multisectorial cooperatives, but they are still headed by a few big ones at the top accounting for most of the cooperative trade (Bijman et al., 2012). The overall trend seems to be an increased degree of concentration, resulting in fewer but bigger cooperatives, and a growing number of transnational ACs with members in more than one country (Bijman et al., 2012).

The EU Common Market has increased food trade among member countries, and, as a result, regional (EU) trade is shaping the increasingly global character of agricultural cooperatives. Many cooperatives have international activities, but the number of transnational cooperatives with members in more than one country is still small (46) (Bijman et al., 2012). Appendix I includes a table published by Copa-Cogeca (2015) with detailed statistics on numbers of ACs, members and turnover across the EU member states.

Agriculture in the European Union: the rise of Producer Organisations (POs)

POs are economic organisations of agricultural producers (or fishermen) with characteristics similar to those of cooperatives. The term is common in the literature on transition countries and development studies (Bijman et al., 2012). A PO may have the legal form of a cooperative, but in many cases it has not. This can be either because the legal requirements for cooperatives pose many restrictions on the activities and the structure of the PO, or, in the case of countries that used to have a socialist state economy, such as Poland or Hungary, because the term cooperative still has negative connotations (Bijman et al., 2012). This is mainly owing to the 'communist legacy', a term used to refer to a distrust of all types of cooperation that ensued from the enforced collectivity of the socialist era and that has discredited most attempts at cooperative endeavours (Bijman et al., 2012).

POs normally focus on new varieties or high-quality (e.g. organic) products and regional specialties (Bijman et al., 2010). Another practical distinction between a cooperative and a PO is that the latter usually has a more focused objective, mainly the joint selling of members' products. Also, a PO is usually positioned in the upstream part of the food chain. Thus, a PO is often involved in joint bargaining with customers, and much less with the processing of members' products. A common definition of a PO accepted by the European Commission is: 'a rural business, owned and controlled by producers, and engaged in collective marketing activities' (Bijman et al., 2012:19). However, the EU legislation explicitly states that a PO can adopt any legal entity or can be clearly defined as part of a legal entity. Thus, organisations other than cooperatives are also recognised as POs.

Even though POs were already included in EU legislation as early as 1972, they became well known (especially in the fruit and vegetables (F&V) sector) when the Common Market Organisation (CMO) for fruit and vegetables was introduced in 1996. The aid scheme for producer organisations was already compatible with the World Trade Organisation (WTO) rules before the 2013 CAP reform (Copa-Cogeca, 2009). The CMO monitors agricultural markets in the EU and covers approximately 90 per cent of EU output (European Commission, 2013c). The number of POs that became members of an association of producer organisations (APO) doubled between 2000 and 2006. The largest APOs are found in Belgium and Italy (Cogeca, 2015).

The EU tries to achieve produce concentration through its policies in order to improve farmers' competitiveness and increase their bargaining power (European Commission, 2014b). In order to access EU subsidies, F&V POs are required to adopt democratic decision-making similar to that of cooperatives. However, in other sectors in which the PO model was introduced at a more recent date, such as hops, olive oil and table olives (for which POs were recognised in 2007) and the dairy sector (POs were recognised in March 2012), POs are only described in terms of their function in increasing farmers' bargaining power, with no mention of ownership or governance models (Bijman et al., 2012). For this reason, POs become a competitor form of organisation, having an impact on the development of the agricultural cooperative movement, as will be discussed when presenting the findings of this research.

POs are normally included in EU reports of agricultural cooperation, even though not all POs are registered cooperatives. Accurate statistics are not available owing to the existence of POs with governance models similar to cooperatives and many hybrids between cooperatives and private companies. In countries such as Belgium and the Netherlands, known for the large, more corporate-like cooperatives, the market share of POs in marketing F&V is higher than that of formal cooperatives (Bijman et al., 2010). In Belgium, for example, the historic "one member, one vote" principle is not mandatory and is followed by only 66 per cent of existing cooperatives, and an enterprise does not need to respect the ICA principles to be considered a cooperative either (Cooperatives Europe, 2016). The legal model is flexible and convenient, and so it is a popular choice for people setting up a business in this country. However, the National Council of Cooperation, an organisation representing only cooperatives that respect the seven cooperative principles, has concluded that, of the 26,626 cooperative enterprises that are currently registered under Belgian law, only 534 are identified as true cooperatives (Cooperatives Europe, 2016).

The main differences between a PO and an AC relate to three areas: (1) the objectives each organisational model is seeking to achieve, (2) the implications of competition law for each, and (3) the legal structure. With regards to the organisational objectives, the EU regulations and member state guidance notes prescribe a more limited list for POs than for ACs: concentrating supply and marketing members' produce, adapting production to the requirements of the market, improving the product and rationalising the mechanisation of production. In the UK, the Rural Payments Agency requests that POs achieve one or more of the following: planning production, improvement of quality, boosting of the value of products, promotion of products, environmental management and crisis prevention (EFFP, 2014).

A second difference to consider is how POs fit into current competition law. In June 2013, the European Commission, the European Parliament and the European Council reached a political agreement on the CAP. One of the key elements of the CAP reform was a significant modification of the application of competition rules in the agricultural sector. The sales of POs in the olive oil, beef and veal, and arable crops sectors were exempted from the application of competition rules. The benefit of this derogation from standard competition rules is only granted if certain conditions are met: the PO should offer members integration of activities other than joint sales, and this integration should be likely to create efficiencies significant enough to offset the possible negative effects of joint selling (European Commission, 2014b). Although ACs are registered as cooperatives (legal forms vary significantly in each member state), the framework in which a PO might be established is not specified, as long as the statute is consistent with its objectives (EFFP, 2014). Therefore, in terms of the legal structure of a PO or AC, there is no difference in how they can, or should, be formed. The effects the different rules for POs and ACs have on the AC sector were brought up by research participants and will be discussed in detail in Chapters 5 and 7.

Finally, producer groups (PGs) are a similar type of farmers' association found in the new member states and in the Mediterranean countries of the EU15 (Greece, Spain, France, Italy and Portugal), where they represented 58 per cent and 40 per cent in 2006, respectively. This type of organisation is used during a transitional period in order to allow PGs to meet the requirements for being recognised as a F&V PO (Cogeca, 2015).

Fruit and vegetables cooperatives in Europe

As discussed in the introduction, this research studied agricultural cooperativism across farming sectors, but with a focus on F&V. The rationale for this decision was based on one of the theoretical assumptions of this book relating to the need for integrated food policy framed in an ecological public health perspective that emphasises the importance of plant-based diets as beneficial to both public health and the environment (Lang et al., 2009; Rayner and Lang, 2012). The F&V sector is of strategic importance and contributes to providing a healthy diet to 500 million European consumers. It currently represents close to 17 per cent of the total value of agricultural produce in the EU and involves approximately 1 million farms (Copa-Cogeca, 2009).

However, the value of production in the EU27 dropped on average by 10.8 per cent for market garden produce and 13.6 per cent for fruit, between 2003 and 2009 (Copa-Cogeca, 2009). Even though Europe is one of the world's largest producers of F&V, alongside India, the EU actually has a deficit and is also the second largest importer (Copa-Cogeca, 2009). As will be discussed in Chapter 5, the effect of multilevel governance is clear in the European F&V

sector: the CAP dictates strict marketing and quality standards, despite the sector receiving hardly any subsidies since the F&V market has been increasingly liberalised (Copa-Cogeca, 2009; Bijman et al., 2012). The almost complete liberalisation of F&V imports to the EU has been formalised in bilateral free trade agreements between the EU and third countries exporting F&V to the EU. Cogeca blames these trade agreements for the rate of exports from the EU being slower than the rate of imports, creating a trade deficit for the majority of fresh produce that went from 7.4 million tonnes (€6.1 billion) in 2002 to 9.8 million tonnes (€8 billion) in 2007 (Copa-Cogeca, 2009).

In 1996, the F&V CMO was transformed, changing its original intervention strategy of using subsidies as defensive instruments into a more proactive approach focused on marketing and rebalancing producers' power (Bijman et al., 2010). Taking advantage of this legislative paradigm shift, many POs emerged in the late 1990s (European Commission, 2014c). This policy shift converted F&V production into a highly regulated but unsupported sector, creating a suitable context for the emergence of POs as organisations with a strong marketing function.

Conclusion

This second chapter has set out the founding ideologies that fomented and influenced the first cooperative experiments in Europe, starting with an overview of the development of agricultural cooperativism on the continent during the eighteenth and nineteenth centuries. The second part of the chapter provided an account of the growth of agricultural cooperation in Europe from the creation of the Common Market until the present, analysing the importance of producer organisations and current figures in key cooperative sectors. The overall trend seems to be increased concentration, resulting in fewer but bigger cooperatives, and a growing number of transnational ACs with members in more than one country (Bijman et al., 2012). The EU Common Market has increased food trade among member countries, and, as a result, regional (EU) trade is shaping the increasingly global character of agricultural cooperatives.

3 Theorising cooperativism and food sustainability

Disciplinary, thematic and chronological streams

The previous chapter introduced the early thinkers of cooperation and the ideologies that gave way to the emergence of the cooperative movement in the nineteenth century. Since then, cooperatives have been the object of study of many disciplines. This chapter consists of two sections that unpack that multidisciplinarity. The first part examines the academic literature that deals with ACs, divided by discipline and thematic stream, including: cooperative behaviours in the biosocial sciences (mainly psychobiology and evolutionary psychology), history, development studies, economic and organisational studies as well as food policy and rural sociology literature. The second section provides a synthesis of the literature around sustainable food systems, discussing the links between sustainability and ACs. At the end of the book, Chapters 10 and 11 link back to the literature review, sharing reflections on the value of a multidisciplinary approach to the study of ACs.

This first section of the chapter explores the five thematic streams identified according to different disciplines, pointing readers to bodies of literature that cover specific dimensions of cooperative studies:

1   biosocial sciences (mainly psychobiology and evolutionary psychology): with a focus on the study of cooperative behaviours;
2   historical stream: history, with a focus on political history of ACs;
3   development stream: ACs seen from a specific socio-economic perspective, mainly in the context of developing countries and emerging economies, but also applied to developed countries; this stream mainly comprises grey literature published by NGOs and the UN;
4   economic and organisational performance stream: two sub-streams were identified here, one of academic literature trying to understand past performance to generate theories that can predict and control future AC performance; the second, grey literature funded by industry and government departments focusing on reporting performance and providing very practical recommendations to improve it;
5   food policy and rural sociology literature: cooperatives as part of the promotion of alternative food networks and shorter supply chains.

Biosocial sciences literature

Even though this stream of literature does not relate directly to agricultural cooperation, it is worth briefly considering some key texts written from an evolutionary psychology perspective, in order to frame the topic of this research within wider debates about the evolutionary and genetic basis of human cooperation. These debates can often be identified as underlying issues permeating all the other disciplines and embedded in the world-view of many practitioners and academics writing about ACs.

Texts from this stream mainly focus on cooperation understood as a construct that refers to a set of behaviours evolved through, and reinforced by natural selection; the main argument in this literature is that survival in the natural world is based, not just on competition, but also on cooperation among individuals, groups and species. When evolutionary thinking permeated other disciplines, such as economics and sociology, two contrasting views emerged: on the one hand, competition was presented as a natural way of surviving in modern capitalist societies; on the other hand, some authors argued cooperation was the basis of (and for) social order. One of the earliest and more popular advocates of cooperation was Prince Kropotkin. The famous Russian activist, scientist and philosopher presented the case for cooperation in his acclaimed books *Mutual Aid* and *The Conquest of Bread*. Kropotkin backed up his arguments with examples from different species that cooperate in the wild to survive (Kropotkin, 1902, 1906). Even though both Darwin and Kropotkin agreed that cooperation helps higher evolution, Kropotkin regretted how Darwin had emphasised the competition element rather than the cooperation, both of which occur in the natural world – both are laws of nature. Kropotkin also analysed the social and political implications of his argument; with regard to rural communities, he wrote:

> The current theory as regards the village community is, that in Western Europe it has died out by a natural death, because the communal possession of the soil was found inconsistent with the modern requirements of agriculture. [. . .] Everywhere, on the contrary, it took the ruling classes several centuries of persistent but not always successful efforts to abolish it and to confiscate the communal lands.
>
> (Kropotkin, 1902:184)

With regard to ACs in particular, Kropotkin thought that they could be both 'anticapitalist' and 'joint-stock individualism' (Kropotkin, 1902:214). The prince accepted that both rural and urban societies struggled to introduce and maintain mutual aid institutions (Kropotkin, 1902:223), but claimed that "the periods when institutions based on the mutual-aid tendency took their greatest development were also the periods of the greatest progress in arts, industry and science" (Kropotkin, 1902:232).

Another key text in this stream is the *Evolution of Cooperation*, a 1981 paper by political scientist Robert Axelrod and evolutionary biologist W. D. Hamilton (Axelrod and Hamilton 1981) that was then expanded in Axelrod's 1984 paper by the same name. The authors applied game theory to analyse how cooperation can emerge and persist, critically reflecting on the implications for – and being informed by – moral and political philosophy of cooperation and the relationship between individuals and groups.

In 1998, philosopher Elliott Sober and biologist David Sloan Wilson published *Unto Others* (Sober and Wilson, 1998), a book using a multidisciplinary approach to demonstrate the evolutionary advantages of altruistic behaviour and group selection. Following Kropotkin's example, this book first provides examples of the animal kingdom and then explores cooperative behaviours in humans. More current literature includes Natalie and Joseph Henrich's book *Why Humans Cooperate* (2007), an anthropological account of cooperation using ethnographic methodology, and evolutionary biologist Martin Nowak's work *Supercooperators* (Nowak and Highfield, 2011). The Henrichs proposed five elements that were key to shaping the evolution of cooperation: kinship (natural selection and cultural learning), reciprocity, reputation, social norms and ethnicity; Nowak's analysis digs down into human bodies, putting forward the theory that cancer develops based on a group of cell's failure to cooperate, opting instead to reproduce in a selfish, uncoordinated manner (Nowak and Highfield, 2011), which has huge implications for treatments.

In summary, this body of literature takes on the mission to uncover the least-well-known aspects of Darwin's theory of evolution that consider cooperative behaviours core to natural selection processes, a topic he expanded on in *The Descent of Man* (Darwin, 1971). This stream argues that the survival values of cooperative behaviours are in line with those of competition, calling for a rethink of socio-economic orders, foreign policies, market strategies and even cancer treatments, as in the case of Nowak's work. This literature challenges mainstream, top–down governance approaches and it is critical of design of public policies and economic systems that is based on competition and the pursuit of individual interest.

Historical literature

This stream includes literature by authors who have written about the historical development of cooperatives in agriculture. A global review of the history won't be attempted, with the focus instead on two examples, Spain and the UK, which are the two country studies that will be discussed in more detail later in the European and global contexts. For detailed reading, the dozens of *Yearbooks of Agricultural Cooperation* that are part of the Plunkett Foundation library's astounding collection are recommended as they offer a fascinating record of the history and development of global agricultural cooperation. These yearly books included statistics on the number, sector and trade of ACs, as well as any other information regarding legislative and political changes that affected the structure of the sector in different countries across the world.

In the case of the UK, the historical study of the cooperative movement focused more on industrial (Webb and Webb, 1897; Coates and Benn, 1976) and consumer/retail (Webb and Webb, 1922; Brazda and Schediwy, 1989; Birchall, 1994) cooperatives, as those were the sectors (especially consumer cooperation) in which UK cooperativism became more successful. This was a reflection of the needs of the working population at that time. Urban industrial workers no longer had the resources to grow their own food, and so cooperation made sense at the retail level to allow them to access food of better quality at cheaper prices.

The most comprehensive historical account of UK agricultural cooperativism is Joseph G. Knapp's book *An Analysis of Agricultural Co-operation in England* (Knapp, 1965). Knapp was the first administrator of the US Department of Agriculture Farmer Cooperative Service. The English Agricultural and Central Cooperative Association commissioned Knapp to write this book-format report owing to his extensive experience in cooperative studies. Knapp was asked to undertake this enquiry into the AC sector because farmers were concerned about the fast degree of vertical integration of food production and the exponential growth of industrial farming (Knapp, 1965). Twenty years after Knapp's report was commissioned, Rayner and Ennew published a historical review of agricultural cooperation in the UK. The authors' account is similar to Knapp's analysis of how the unfavourable policy context of the 1920s and 1930s hindered the development of ACs in the UK; they also highlighted how the incorporation of Britain into the EU was the cause of a shift in policy and a new emphasis on promoting agricultural cooperation, resulting in its substantial expansion, aided by UK and European funding (Rayner and Ennew, 1987).

Going back to the Plunkett Foundation's *Yearbooks of Agricultural Cooperation*, these fascinating reports included specific country chapters, providing an excellent account of the gradual industrialisation of agriculture and its socio-economic and political context and impacts that emerged (a good example of this rich description can be found in the *Yearbook of Agricultural Cooperation 1966*, Plunkett Foundation, 1966). For example, numerous historical-political and political-historical accounts of Spanish agricultural cooperativism were found in these year books. Historical politics, referring to the instrumental uses of history, is a concept used to reflect how historical narratives can be intentionally reconstructed and disseminated, not only within a country, but also across national boundaries, frequently with a political motive (Miller and Lipman, 2012). The historical connotations attached to cooperatives not only shaped how the origins and development of the movement were and are perceived by the general public, but also affected policy approaches to ACs and the dynamics of collective memory. Examples of this use of history records were found while reviewing the Plunkett *Yearbooks of Agricultural Cooperation* that included a chapter on Spain (1938, 1954, 1955, 1961, 1966, 1967, 1973, 1985 and 1988). The chapters were written by different authors every year, sometimes by Spanish authors (e.g. the *Yearbooks of Agricultural Cooperation 1961* and *1967*; Plunkett Foundation, 1961, 1967), sometimes by English reporters (*Yearbook of Agricultural Cooperation 1985*; Plunkett Foundation, 1985), with mainly single contributions in any given year. The structure of the texts varies from year to year: some chapters provide statistics, whereas others analyse case studies of specific regions. It was interesting to notice how the ideologies of the authors are quite palpable in the texts, and how the accounts of agricultural cooperativism change accordingly. The chapter on Spain included in the yearbook for 1967, written by Jose Luis del Arco Alvarez, offers a good example to illustrate this point. According to Del Arco, the development of cooperation was almost exclusively achieved by Catholic social action groups, many years before any legislation was introduced; the author also adds that the lobby pressure of these groups resulted in the passing of the Law on Agricultural Syndicates in 1906. This historical account embellishes the role the Church played in promoting conservative ACs that maintained the status quo, a part of Spanish ACs' history that many other authors have highlighted (Garrido Herrero, 2003; Simpson, 2003; Hermi Zaar, 2010; Planas and Valls-Junyent, 2011).

The chapter of the *Yearbook of Agricultural Cooperation 1954* authored anonymously by "A Correspondent" had a completely different political tone. This alternative account explains how the National Federation of Consumer Cooperatives and the Cooperative Supply Centre set up before the Civil War were regarded by the end of the war as 'Red' and communist, and thus were not further developed. The author also blames the "peculiar structure" of cooperatives during Franco's dictatorship – characterised by compulsory vertical integration and the obligatory presence of a representative of the regime and the Church on each board – as the reason for the lack of links with cooperatives in other countries (Plunkett Foundation, 1954). An additional example can be found in the 1953 book's chapter on Spain, which mentions how the Franco regime's law of 1938 was designed to annul the socialist characteristics of the previous cooperative law approved during the Second Republic, and compares the legislative framework of the regime to that of fascist Italy and Pétain's France (Plunkett Foundation, 1953).

With a strong political element, a more recent stream of academic papers developed in Spain after the end of Franco's regime; the return to democracy brought a resurgence of literature trying to study the historical development of agricultural cooperativism, creating a heated debate, still alive today, on the roles that the Second Republic, Franco's Regime and the Church had in it (Bernal et al., 2001). Readers interested in this stream are recommended to read Garrido Herrero's work (Garrido Herrero, 2003). Some of the more recent historical literature links early agricultural cooperatives to notions of social capital, food democracy and peasant rights (Planas and Valls-Junyent, 2011; Tapia, 2012). There is confusion over terms in the literature, and many authors use *cooperativas* and *asociaciones* interchangeably in the same paper, as discussed at the beginning of this chapter. It has been highlighted that the early history of Spanish cooperativism has not been reported and documented accurately from its real beginnings owing to the confusion over these terms (Plunkett Foundation, 1967).

Some authors (Planas Maresma, 2008) have analysed farmer cooperatives within the wider frame of agricultural associationism, presenting cooperation as a struggle to maintain different collective identities in Spain's increasingly urbanising society at the beginning of the twentieth century. Planas Maresma's political analysis of agrarian cooperativism uses the lens of collective identity in Catalonia. According to Planas Maresma, the dual character of cooperativism at the time, with co-existing 'rich' and 'poor' cooperatives that served different purposes for different actors, weakened the movement (Planas Maresma, 2008). Other authors have tried to pinpoint reasons for the numerous but short-lived ACs that developed at the beginning of the twentieth century in Spain. Many conflicting accounts of the power relations in agrarian cooperation in this troubled period of Spanish history still exist (Simpson, 2003; Garrido Herrero, 2007; Planas and Valls-Junyent, 2011).

Finally, it is interesting to share here that only one academic paper was found that was written in English and had a historical EU/international focus on AC history. The article, written by Morales Gutierrez and colleagues (2005), offers a comparative review of historical records of agricultural cooperative movements during the twentieth century in seven European countries (Belgium, Denmark, France, Italy, the Netherlands, Portugal and the United Kingdom). The authors compared the legal framework, state support for the promotion and development of cooperatives, the underlying values involved and the different sociopolitical and economic factors that fostered or curtailed the development of ACs in these countries (Morales Gutierrez et al., 2005). Based on their findings, the authors identified two opposing development strategies for agricultural cooperation in the selected case studies based on two dimensions: the level of ideological inspiration and the degree of support from the state (Morales Gutierrez et al., 2005). These dimensions are still relevant to ACs today, as will be discussed in Chapters 5, 10 and 11.

Development studies literature

This more recent thematic stream emphasises the wider socio-economic benefits associated with AC enterprises. The texts reviewed in this section are in the majority grey literature coupled with some academic publications; the focus is on both developed and developing countries, but from a different approach in each case. When focusing on developed economies, a dual understanding of development was identified:

- Developing rural areas/communities: the most active publisher in this area is the Plunkett Foundation, the aforementioned charity currently working on facilitating community-owned farm projects. Also, in 2013, Co-operatives UK published a paper on *sticky money*, referring to how, for every £1 that was spent with and by a cooperative, the local economy received £1.40. The study followed Local Multiplier 3 (LM3) methodology, a long-established way of measuring how money travels in local economies (Sacks, 2013). Additionally, some of the academic literature reviewed considered cooperatives' potential to develop tourism in rural areas (Barke and Eden, 2001) and generate social cohesion (Fairbairn, 2006) and community development (Zeuli and Radel, 2005).

- Developing alternatives to industrial farming and capitalism: this includes texts from the coordinators of the Campaign for Real Farming (Tudge, 2011) and the New Economics Foundation (NEF, 2011) that indirectly discuss the social and financial potential of the cooperative model.

In low-income countries, many NGOs and UN publications also present ACs as a tool for development, poverty reduction and combating gender inequalities (e.g. Fairtrade Foundation, 2011; FAO, 2012; Oxfam, 2012). Other authors, such as Birchall, have described how ACs have helped small farmers in developing countries reach export markets (Birchall, 2003), while Rhodes (2012) has highlighted the paternalistic, Eurocentric character of this approach. As this research is focusing on EU countries with industrialised agricultural sectors, this last development stream was only briefly explored to frame the object of study within a global perspective. However, the data collected by this research highlighted the extent of the impacts European ACs have in developing countries, a topic that will be discussed in detail in Chapter 10. Finally, other global organisations, such as the International Labour Organization (ILO), also regularly commission research and publications on cooperatives, as well as guidelines for cooperative legislation and recommended levels/areas of state intervention in cooperative affairs (Henrÿ, 2012).

Economics literature

The largest body of academic literature on ACs comes from the economics discipline. This section also reviews recent grey literature from industry bodies. The review showed a marked interest in ACs from economists in the 1980s. As will be discussed, this was the decade when ACs underwent a process of transformation to become 'more professional food businesses' (Chaddad, 2009). This transformation was both fostered and documented by the economics discipline. This stream also included abundant industry data providing statistics on and forecasts of ACs' performance, such as publications from the European Commission, Co-operatives UK, Copa-Cogeca and Cooperativas Agro-alimentarias, which are referenced throughout the book.

However, before discussing more recent literature, it is interesting to refer back to one of the most famous political economists, Karl Marx. Chapter 2 introduced Marx's views on the topic: although he thought cooperatives were not radical enough, he considered they were a step in the right direction towards the "great workers' companies" he envisioned. In 1866, at the Provisional General Council of the International Working Men's Association, Marx encouraged workers to get involved in production cooperation, rather than consumer cooperatives, as production and industry "attack the groundwork" of the capitalist system (Marx, 1866). From these early Marxist and Weberian discussions around whether cooperatives could bring down the capitalist class system or not, the debate in economics gradually moved to the other end of the spectrum: how to make ACs as flexible (in terms of capital accumulation and governance structures) as privately owned firms (POFs) in order to compete in an increasingly concentrated market.

In 1947, two US petroleum business associations commissioned a report to the Austrian economist von Mises on the role of cooperatives in the petroleum industry, concluding that the organisational model of cooperatives was a failure and only existed thanks to governments' support. Despite the shortcomings of this argument, which ignores the support and indirect subsidies that many private corporations receive from the state, it is worth including a quote from von Mises regarding his views on agricultural cooperatives:

> The farmers are producers. But the farmers' cooperatives do not organise the farmers in their capacity as agricultural producers; they organise the farmers only as buyers of various equipment and articles required for their production and as sellers of the products. The individual farmer remains an independent entrepreneur and is, as far as his production activities are concerned, not integrated into a cooperative production outfit.
>
> (von Mises, 1990:239)

The above quote reflects the apparently contradictory identities that farmer members of ACs often struggle with: growers, buyers of inputs, consumers of food, and so on. This multiplicity of identities and ACs' failure to organise farmers as workers will be discussed in Chapters 8, 9 and 10.

Historically, in a theoretical effort to fit ACs within one of the generic forms of economic organisation, many authors applied economic theories to the study of cooperatives. In 1962, Helmberger and Hoos published the first mathematical model of an AC based on the neoclassical economic paradigm. Soon a debate started in the economics community around how to define cooperatives from a new institutional economics (NIE) perspective that could take into account the social costs of doing business, conceptualised in an analysis of transaction costs and property rights (Karantininis and Nilsson, 2007). Much has been written about the high transaction costs of cooperative governance, mainly from this NIE perspective (Karantininis and Nilsson, 2007). It is often claimed in the literature that such a complex governance structure is bound to make the organisation fail or convert to a private company (Lindsay and Hems, 2004; Münkner, 2004), an organisational theory labelled the "degeneration thesis"(Somerville, 2007).

The bulk of these papers have a common denominator: cooperatives are perceived as an inefficient organisational model, not fit to survive in a competitive market owing to its costly democratic decision-making process, risk-averse nature and long-term investment preferences. The main objectives of these papers can be summarised as: (a) to identify the model's weaknesses and (b) to recommend strategies to overcome these weaknesses (Murray, 1983; Oustapassidis, 1988; Ortmann and King, 2007). Some authors have championed an opposite way of thinking that tries to prove and provide evidence for the resilience of the cooperative model (Ostrom, 1990; Sanchez and Roelants, 2011). For a detailed account of advances in economics cooperative theory since the 1990s, the work of Cook and colleagues is recommended (Cook et al., 2004).

How to define the economic organisational form of ACs has been the topic of lively debate in this body of literature. The extract below from Chaddad's paper paints a clear picture of the evolution of the perceived organisational identity of cooperatives:

> Until the early 1960s much of the theoretical debate focused on whether cooperatives represented a form of vertical integration by farmers, that is, an extension of the member firms [. . .] or whether cooperatives could legitimately be analysed as organisations having scope for decision making independent of their member firms [. . .] Subsequent theoretical work has modelled cooperatives based on coalition [. . .] nexus of contracts [. . .] and property rights perspectives [. . .] There is, however, no consensus yet as to how to define a cooperative in the continuum of generic forms of economic organisation. [. . .] I show that cooperatives blend market-like attributes with hierarchy-like mechanisms and thus should be viewed as a true hybrid rather than as an intermediate form.
>
> (Chaddad, 2009:1)

Four thematic and chronological streams were identified during the review. The first stream emerged in the late 1990s and reflected a shift in the way ACs were studied and portrayed in the literature. During the late 1990s and early years of the twenty-first century, a large body of work on ACs focusing on analysing the "new generation cooperatives" (NGCs) was published. NGCs started to emerge in the US during the late 1980s as a survival strategy to overcome the aforementioned crisis they were encountering by adopting more liberalised models of governance and investment. Table 3.1 compares the main differences between traditional cooperatives and NGCs (based on Merrett and Walzer, 2004, and Cook and Plunkett, 2006). Whereas traditional cooperatives focused on production and countervailing market power, NGCs emerged as a more entrepreneurial model, more market-oriented and with a focus on value added, normally taking the cooperative activities downstream within the food chain (Merrett and Walzer, 2004). Stofferahn has written about how the conversion of ACs to NGCs snowballed and was presented as the only sensible option for farmers, dissipating any resistance from cooperative members against the privatisation of their cooperatives (Stofferahn, 2010). Other authors questioned whether NGCs represented a new type of cooperative or the beginning of the end of the cooperative structure itself (Ménard and Klein, 2004).

Table 3.1 Comparing the main differences between traditional and new generation cooperatives

	Traditional	Entrepreneur (NGCs?)
Main objective	Achieve economies of scale	Horizontal competition with other producer companies/vertical competition with retailers
Strategy	Operational	Strategic
Focus of value-added level	Production/bargaining/ low processing	Processing of high yields/ marketing/branding
Portfolio	Single commodity	More varied to become/ remain preferred supplier
Membership	More homogeneous	More heterogeneous (but still reliant on members' commitment)
Financial structure	Members' capital	Differential levels of openness to external shareholders
Character	More local	Differential levels of internationalisation
Relationship focus	Transactional relation	Capital relation
Shareholders	Members only (farmers)	Members and non-member investors with differential degrees of decision making
Preferred growth method	Organic growth (increased production and members) Horizontal mergers	Vertical integration Internationalisation
Intangible value added	Social and potentially environmental by localisation and farming method	Branding Potentially environmental by larger equity to fund technology for precision farming
Focus of collective action	Defensive = protect position in failing market	Offensive
Products traded	Undifferentiated commodities	Branded products
Level of capital accumulation	Defensive = member level	Offensive = coop level
Capital intensity	Less capital intensive	More capital intensive

Source: Adapted by author from Merrett and Walzer (2004) and Cook and Plunkett (2006)

At the turn of the twentieth century, several authors proposed the US NGC model as the way forward for European ACs. Gradually, reflecting the increasing complexity of AC models in terms of ownership, economists' efforts focused on developing typologies of ACs based on property rights and governance structures (see Harris et al., 1996; van Bekkum and van Dijk, 1997; Cook and Iliopoulos, 1999; Nilsson, 1999; Fulton, 2001; van Bekkum, 2001; Goldsmith and Kane, 2002; Iliopoulos, 2005). As discussed in Chapter 1, many typologies emerged trying to categorise ACs based on the source of their investments (from private investors or farmer members). Other authors theorised the existence of organisational life cycles, suggesting that, as ACs grow and mature, the managers' aspirations take over members' ones, gradually transforming the organisational and financial structures and trading practices and making them more akin to investor-oriented firms than traditional cooperatives (Hind, 1999).

As part of this interest in cooperative identity, several authors explored extensively (from an economic perspective) the effects a flexible cooperative identity has on cooperative principles and cooperative structures when competing in a liberal market context. Szabó has talked about changing the "cooperative identity" to generate a self-defence mechanism against corporations by becoming more like them (Szabó, 2006); other scholars have emphasised ACs' role in the social economy, helping the rural economy to adapt to the European policy context (Juliá Igual and Marí Vidal, 2002). Kalmi has tried to explain the disappearance of the cooperative model from economic textbooks by putting forward the argument of a paradigm shift from institutional to neo-classical analysis (Kalmi, 2007). Other authors have focused on trying to prove the theoretical benefits of defining ACs as "hybrids" (Szabó, 2006; Ortmann and King, 2007; Chaddad, 2009), including recognising their multiple objectives beyond profit-making for members.

From 2007 onwards, perhaps caused by the effects of the financial crisis, the literature reflects a concern with the increasing desocialisation of ACs (suggesting many are becoming more capital-centred rather than people-centred), while emphasising and reminding readers of the importance of social capital (Somerville, 2007; Gómez López, 2009; Bijman et al., 2010; Feng et al., 2011; Mulqueen, 2011; Whyman, 2012; Nilsson et al., 2012; Jones and Kalmi, 2012). Interestingly, this is in contrast with the growing popularity of buying groups and other types of informal consumer cooperative, as well as new attempts to resist the monopoly of supermarkets and link up consumers with small farmers (Kneafsey et al., 2008; Böhm et al., 2014). A more critical perspective on ACs comes from Jacques Berthelot, a French agri-economist working for the NGO Solidarité, who in 2012 published a daring piece titled "The European agricultural cooperatives, promoters of the unequal globalisation". In this article, Berthelot uncovers the bad practices of some large European ACs, including the uncooperative conditions French sugar beet AC Tereos imposes on its subsidiaries in Brazil and Mozambique. Other cases relate to land grabbing and outsourcing of operations to developing countries, where some ACs exploit workers who contribute to the commercial activity of the AC but who are not given the option to join as members (Berthelot, 2012). Berthelot's work is referenced again in Chapter 10 when considering the global impacts ACs have beyond their national boundaries and how their practices often clash with cooperative principles.

Most of the academic literature from the economics discipline is mainly concerned with an organisational level of analysis. This narrow focus was considered a limited analytical lens for this research, as the present study was informed by and applied a food policy perspective that aimed to look at ACs as part of the food system, including how they are helping or hindering food sustainability and how they fit within the wider cooperative movement and other social movements.

Food policy literature

Finally, after discussing some of the main disciplines that have considered cooperatives a topic worthy of study, this section offers a review of the body of literature specifically focused on food policy. Before continuing, it is worth pausing to explicitly specify the definition of food policy that was used as the backbone of this research: "The decision-making that shapes the way the world of food operates and is controlled" (Lang and Heasman, 2015:2). For the purpose of this study, food policy was understood as a comprehensive concept that refers, not only to the way the different levels of the food system are organised and governed, but also to the policy-making process itself, and how it is shaped and influenced by a variety of actors with conflicting interests and expectations (Lang et al., 2009). Food policy has a multidisciplinary approach that tries to unravel the dynamics of the increasingly globalised and industrialised food system. It is essential to highlight the interdisciplinary and contested character of food policy and to point out that a lack of policies in some unregulated areas also affects what people eat, either by deliberate or unintended causes and effects; thus, unregulated areas of the food system are also part of the remit of food policy study in general, and this book in particular. This definition informed the selection of academic literature reviewed in this stream, including work from scholars who define themselves as rural sociologists and study the policies and politics of the AC sector.

The search criteria excluded studies on developing countries, as this book is focusing on the EU context, although, as mentioned in the previous section, Chapter 10 also considers the impacts European ACs have globally. The majority of literature on cooperatives identified related to consumer cooperatives rather than ACs. This finding could be explained by the current emphasis on consumer behaviour and individualistic rather than systemic approaches as the entry point to change harmful dietary trends (Lang and Heasman, 2015). Clashing interests that hinder policy integration dominate the contested food policy terrain. In this context, placing the responsibility on the consumer rather than on the whole food system is an ineffective but easy approach from which powerful actors benefit (Lang et al., 2009). It also reflects how an ever-decreasing farming population becomes relegated by the study of an ever-increasing urban consumer population. When ACs do appear in this body of literature, they do so as part of shorter food chains and AFN discourses (Ronco, 1974; Allen, 2004; Lamine, 2005; Guzmán and Mielgo, 2007; Hingley et al., 2011). The most common themes identified included: arguments to present the cooperative model as a tool to create new, alternative ways of buying or producing food (Guzmán and Mielgo 2007; Little et al., 2010; Calle Collado and Gallar, 2010; Calle Collado et al., 2012); and ACs as a tool to foster rural development, with a clear interest in analysing case studies of small (both in terms of trade and size) ACs that are usually part of niche markets (Seyfang, 2006; Moragues-Faus and Sonnino, 2012).

Overall, the literature on ACs is mainly reductionist, as it tends to be strictly either economic or sociological (the latter in very few cases), ignoring the ongoing tensions, dilemmas and contradictions in the movement (Gray, 2014a). Very few authors have written about ACs in relation to the neo-liberal and political context in which they exist. The main work in this area that this book draws on it is that of rural sociologists Patrick Mooney and Thomas Gray. Mooney and colleagues' thought-provoking papers from 1995, 1996 and 2004 discussed the increasing depoliticisation of ACs taking place through the reprivatisation discourse of neoclassical economics, pushing ACs to achieve a higher degree of financial competitiveness by becoming more capital-centred rather than people-centred (Mooney and Majka, 1995; Mooney et al., 1996; Mooney, 2004). Gray et al. (2001) discussed how very few ACs were suited to compete with agri-food giants. Those cooperatives that were able to enter into competition with multinationals did so at the cost of shifting towards positions that were "characteristically less cooperative, and more bureaucratic, and more top down, though likely more efficient, and with greater market penetration" (Gray et al., 2001:167).

Mooney proposed to reclaim ACs as organisations that have the potential to legitimise and sustain class struggles, tackle power imbalances and improve workers' conditions (Mooney, 2004). In agriculture, cooperatives can help farmers regain power from supermarkets by strengthening their negotiating position (Mooney, 2004). In his view, cooperatives can help farmers regain power in the food system while (and because of) raising contradictions at different levels: (1) in social relations, between production and consumption; (2) in spatial relations, between the local and the global; and (3) in collective action, between cooperatives as both traditional as well as new social movements. In this sense, Mooney notices that ACs provide the space for those tensions to emerge, become visible and provide innovative solutions, in contrast to the neoclassical economics model that presents those tensions as barriers to efficient profit-making (Mooney, 2004:81).

Few authors since then have written about the increasing "corporatisation" of agricultural cooperatives. In their study of cooperatives in Scotland, Wilson and MacLean concluded that the cooperative members they interviewed had a "collective belief in independence" rather than any bigger aspirations to transform the market economy status quo and were "motivated primarily by individualism and survival rather than shared or cooperative values" (Wilson and MacLean, 2012:535). The authors found that the word "cooperative" is often not used in ACs' names because "a lot of co-ops have tried not to look like co-ops", as the cooperative identity is seen as "old-fashioned" (Wilson and MacLean, 2012). A popular alternative to the word cooperative is "farmer-owned business", a term that recurs in the literature and many ACs' websites, as discussed at the beginning of this chapter.

ACs have also been touched upon in discussions of contemporary peasant farming (van der Ploeg, 2013). In his book *Peasants and the Art of Farming. A Chayanovian Manifesto*, van der Ploeg refers to the evolution of ACs from Chayanov's days (the beginning of twentieth century) when cooperatives were "class based" and "still offered the promise of an effective countervailing power" to the present: *"(t)oday the situation is very different. Former cooperatives have evolved into entities that treat peasants in the same way as food empires"* (van der Ploeg, 2013:84). The work of van der Ploeg is discussed in more detail in Chapter 4 as part of the theoretical framework of this research.

Two more recent papers, by Emery (2015) and Stock et al. (2014), have identified the relationship between neo-liberal conceptions of autonomy, independence and their effects on agricultural cooperation. In turn, Gray has discussed how many US ACs are becoming large bureaucratic organisations in their own right, shifting cooperative governance models away from economic democracy and towards IOF models that emphasise the needs of capital over members' needs (Gray, 2014b). The author has warned how, when a return-on-investment logic is used, the traditional local embeddedness of ACs becomes an unnecessary constraint that interferes with mobility of capital. As a result, many ACs have expanded geographically, losing their local identity and uniqueness and adding "another layer of distance – physical distance – between members, member governance and cooperative decision-making" (Gray, 2014b:26). These concerns will be discussed in more detail in Chapter 5.

In 2017, the *Journal for Rural Studies* published a special issue on "The more-than-economic dimensions of cooperation in food production", with a focus on conventional agricultural cooperatives and the complexity of processes and values included in cooperation (Emery et al., 2017). In this special issue, several authors consider the ongoing analysis of tensions between individual and collective interests in ACs (Emery et al., 2017; Forney and Häberli, 2017; Wynne-Jones, 2017). In this issue, I discussed how multi-stakeholder cooperatives with a diverse membership of growers, workers and consumers can help cooperatives remain true to their principles and members through direct accountability and having to accommodate different interests, rather than ignoring them.

Forney and Häberli argue in this issue against reductionist dichotomies and simplified notions of hybridisation between two poles of "traditional" and "corporatised" cooperatives as they can fail to reveal the multiplicity of motivations different members might bring to the table and help produce and reproduce in their cooperatives. I agree that avoiding dichotomies is a very worthwhile approach to move the analysis of ACs forward and to work towards the shared vision that Fonte and Cucco (2017) proposed for an ideal future in which cooperatives operating in alternative food networks and well-established farmers' organisations join efforts and activities. Nevertheless, this book argues that there has not been enough research to unpack how and in what ways ACs are moving away from being a transformative movement to being reduced to a tool to survive market developments. This research opts for a solidary-critical approach (Favaro, 2017) through which to acknowledge and appreciate the important role ACs play in supporting individual farmer members, while still exploring difficult issues around co-optation that reveal a somehow uncomfortable, but potentially fruitful, analysis to help ACs move towards a fairer and more sustainable future. To end this section and introduce the next discussion on the challenge of conceptualising food sustainability, I list below three reasons why it is important, not only to highlight the tensions between ACs' practices and cooperatives' survival, but to do so from different perspectives:

1   From a consumer perspective, there is a need to provide a fair, realistic picture of ACs for consumers. Consumers perceive ACs in a positive manner, as they assume farmers' cooperatives are more ethical and value-led than private companies (Cooperativas Agro-alimentarias, 2013). There is a responsibility to provide an accurate picture of ACs' practices, rather than involve them in a cooperative halo effect.
2   From a state/policymaker perspective, there is a need to inform the best approach to foster cooperation among farmers in order to design effective subsidy policies that do not direct public funds to farmers' organisations that might be reproducing the wrong kind of unequal cooperation, as will be discussed later with reference to the increasing importance of POs in the EU and the impact some ACs' practices have in developing countries (Berthelot, 2012).
3   From an AC perspective, the ruthlessness of the market and agri-food industry forces ACs face also needs to be taken into account, since these shape the room for manoeuvre ACs have in their practices and business models. How can farmers be supported to survive these market and industry forces without losing their voice as members, and without risking the environmental sustainability and viability of their farms?

The challenge of defining sustainable food systems

A growing amount of evidence of the environmental challenges facing the food system has been piling up for decades (Pimentel et al., 1973; FAO, 1995; McMichael et al., 2007; Garnett et al., 2013; IPCC, 2014). Lang has collated evidence from other authors to neatly summarise these challenges in what he calls the New Fundamentals (adapted from Lang, 2010, with additional and more up-to-date statistics added by the author from several sources referenced accordingly):

- **Climate change**: agriculture is both a perpetrator and a victim of climate change; food production accounts for 20–30 per cent of anthropogenic greenhouse gas (GHG) emissions (Garnett, 2014), fuelling climate change that in turn is creating more extreme weather events that are affecting global yields (IPCC, 2014). Of these emissions, animals are responsible for approximately 31 per cent and fertilisers for 38 per cent (Stern, 2006). In Europe, food is consumers' biggest source of GHG emissions (Tukker et al., 2006, 2009). The consultancy firm KPMG found that, if the global food production sector had to pay the full costs of its environmental impact in 2010, it would have reported a loss of 224 per cent (KPMG, 2012). More recently, the Sustainable Food Trust has estimated the hidden cost of food in the UK to be 100 per cent of the cost consumers pay for food, so that, for every pound spent, consumers pay another pound through other channels to deal with non-accounted costs such as cleaning up rivers or health costs for dealing with diet-related diseases (Sustainable Food Trust, 2017).
- **Energy**: This especially refers to reliance on non-renewable fossil fuels and increasing demand for production of agrochemical inputs, operating machinery and transport to cover growing distances between farms, retailers and consumers (UNEP, 2009).
- **Loss of agricultural biodiversity**: During the twentieth century, an estimated 75 per cent of the genetic diversity of domestic agricultural crops inherited from the nineteenth century was lost (FAO, 1995). Nowadays, 60 per cent of the calories and 56 per cent of the protein derived from plants for human consumption come from only three crops: rice, wheat and corn (Thrupp, 2000; Khoury et al., 2014). By the end of the twentieth century, 12 plant species accounted for 75 per cent of the global food supply, and only 15 mammal and bird species accounted for 90 per cent of animal agriculture (FAO, 1998; FAO, 2010b; Khoury et al., 2014).
- **Waste**: A report from the House of Lords for the European Committee (2014) calculated that EU food waste equals 89 million tonnes per year (one-third of all produced food), almost as much as the entire net food production of sub-Saharan Africa (House of Lords EU Committee, 2014).
- **Labour**: severe labour shortages are predicted, as food-related work is not properly recognised or fairly paid, reflected in unbalanced financial flows; in the UK for example, farming accounts for only 8 per cent of the total gross value added of the agri-food sector (DEFRA, 2012).
- **Demographics**: this includes several factors such as population growth, urbanisation and inequity of consumption rates. In European farms, the trend is clear: one-third of farmers are over 65 years old, and only 7 per cent are under 35 (European Commission, 2013a).
- **Land**: tensions around land use are becoming more acute, for example, between food and biofuels production, but also regarding cases of land grabs,

which are increasing food security concerns (UNEP, 2009). Additionally, even though urban areas only cover 3 per cent of the Earth's surface (Scheneider et al., 2010), their impact through trading and consumption is huge, with ecosystem support needs (including waste absorption) sometimes 500–1,000 times larger than their own area (Folke et al., 1997).

- **Water**: agriculture accounts for 70 per cent of potable water use, livestock being a significant part of it (WWF, 2013) while at the same time also being a major source of water pollution. Intensive livestock production is thought to be one of the largest sector-specific sources of water pollution (UN, 2011).
- **Soil:** with increasing erosion and degradation due to intensive farming and livestock production, the loss of soil is an increasing concern (Garnett, 2014). Fertilisers and intensive farming have also dramatically altered nitrogen, phosphorus and carbon cycles (Rockström et al., 2009a, 2009b; Cordell et al., 2009; Foley et al., 2011).
- **Health/nutrition transition**: this refers to the trend that reflects the dietary change of populations in emerging economies, with an increase in high protein intake (especially of animal origin) and sugary foods (Popkin, 2003). By 2008, more than 1.4 billion adults, 20 years old and above, were overweight (WHO, 2013). A double burden of health problems from over-, under- and mal-consumption and non-communicable diseases now co-exist in most countries across the world (WHO, 2011).

Additionally, there has been growing concern over the environmental impact of meat production, an issue that cuts across all the above fundamentals. The FAO took a major role in raising awareness about this problem with its 2006 Livestock's Long Shadow report (Steinfeld et al., 2006). Approximately half of all cereals grown globally are fed to animals destined for human consumption, and they account for a tremendous amount of embedded water (Steinfeld et al., 2006; Gerber et al., 2013; Lymbery and Oakeshott, 2014). For example, it takes 2,400 litres of water to produce a 150-gram hamburger (Chapagain and Hoekstra, 2006). Therefore, increasing calls for a reduction in meat consumption is taking place, not only owing to its environmental impact, but also for animal welfare and public health reasons (Lymbery and Oakeshott, 2014).

These externalised costs are a huge burden on society, public health and ecosystems (Lang et al., 2009; Mason and Lang, 2017). We are dangerously altering biophysical processes, threatening the ability of Planet Earth to sustain life as we know it, a fact captured in the concept of "planetary boundaries" (Rockström et al., 2009a, 2009b). Many academics and civil society organisations have been calling for policy change and system transformation for years, highlighting that some planetary boundaries have already been exceeded and others are approaching hazardous limits (Rockström et al., 2009a, 2009b; Steffen et al., 2015).

For all the above reasons, environmental sustainability is one of the most important aspects of the crisis facing the current food system and also an essential feature and aim of an integrated approach to food policy (Lang et al., 2009). Attempts to define sustainability started back in the late 1980s with the popular definition of sustainable development (SD) offered by the Brundtland report, in which SD is defined as the "development which meets the needs of the present without compromising the ability of future generations to meet their own needs" (WCED, 1987:8). However, this definition was framed by a conception of sustainability still based on constant economic growth. This conception of SD fosters a continuation of incongruous policies to secure growth that damage the environment, and thus it compromises the actual type of development it is calling for (Frazier, 1997). For this reason, when using the term sustainability, it is more constructive to refer to the UNDP's comprehensive conceptualisation of environmental sustainability and human development that goes beyond "beyond income and growth to cover the full flourishing of all human capabilities" (UNDP, 1996:49). However, these earlier definitions of sustainability still do not cover the complexity that food represents. Food is the link between farming methods and nutritional needs, while being interwoven in deeply embedded cultural traditions and new fashions that transcend any attempts to reduce sustainability to GHG emissions or food miles.

It was the year before the Brundtland report was published that the notion of sustainable diets first appeared in the academic literature (Gussow and Clancy, 1986), highlighting how human nutrition is intrinsically linked to natural ecosystems. In their paper, Gussow and Clancy argue that, "in our time, educated consumers need to make food choices that not only enhance their own health, but also contribute to the protection of our natural resources" (1986:1).

Early movements for sustainable food systems and sustainable farming were concerned with the environment but also with social justice (Allen, 2004). These comprehensive understandings of sustainable diets considered two main factors. In development policy, the focus was on food security, whereas in affluent countries, the link between nutrition and the environment was diluted over time as the constant availability of cheap food meant that the social justice element became less relevant, and lobbying efforts went down the environmental route (Allen, 2004; Lamine et al., 2017). The debate became centred on farmers (level of actor), agriculture (level of food system) and the environment (impacted area). In the 1990s, the concept was soon criticised for ignoring issues of social justice, and critics called for a wider conception of sustainability, extending the sustainable agenda beyond the farm gate and bringing consumers into the debate (Kloppenburg et al., 2000). However, this new breadth of aspiration generated a multiplicity of definitions; even today, there is no commonly accepted definition of what a fair and sustainable food system (SFS) should look like. As an example, in a study analysing the evolution of definitions of alternative food systems, Allen found that different NGOs and alternative agri-food institutions defined social justice – a key ingredient of an SFS – in quite different ways (Allen, 2004).

Some of the earliest attempts to map the attributes of a SFS started in the late 1990s, following the publication of the Brundtland report. In 1997,

Clancy and Lockeretz (1987) called for a food system that included a more agriculturally literate society, local food security, and supportive institutions and policies. In 2000, Kloppenburg and colleagues asked 125 people from the alternative food scene (working in farming and food projects) to list the attributes they thought an SFS should have. The group came up with a long list of attributes. In the paper, the authors highlight how the disputed nature of the concept of sustainability has facilitated opportunistic agri-industry attempts to appropriate the term for their own purposes. As the authors suggest, the lack of clarity has diluted the meaning and power of this concept (Kloppenburg et al., 2000) and it has allowed "the conventional Green Revolution experts to sell their old wine in the new bottle of 'sustainable agriculture'" (Lélé, 1991:617). The concept of *"sustainable intensification"*, discussed below, is an example of this co-optation.

Kloppenburg and colleagues offered another example to prove this point: in 1991, DuPont Corporation defined an SFS as "ecologically sound, economically viable, socially acceptable" (Kloppenburg et al., 2000:179), but, as the authors highlight, "what is socially acceptable is not necessarily just" (Kloppenburg et al., 2000:179). This example emphasises the importance of using food democracy as a key guiding principle of SFSs. Apart from mapping the complexity of conceptualising an SFS, one of the key contributions of Kloppenburg's paper is the emphasis on the importance of integrating producers and other actors into the definition process, as they are the key agents of change:

> Conceptual framings of alternative food systems have been based largely on the reflection of academics and policy specialists. While such formulations are surely valuable, they may not reflect the full range of understandings characteristic of the producers and eaters.
>
> (Kloppenburg et al., 2000:178)

This democratic aspect of the defining process will permeate through the present book as it aims to collate ACs and ACs' members' conceptions of SFSs.

Allen has reported how nowadays most attempts to define SFSs use permutations of the three "Es" model: E for environment, economics and equity (Allen, 2004:82). However, Allen warned about the common underlying limitation of these models and definitions of sustainability and thus of SFSs: the ambition to maintain the system for current generations with the unequal distribution of both food and environmental problems that we have today, instead of aspiring to construct a better system now (Allen, 2004). Allen argues that this focus on future generations ignores the unmet needs of people living in the present. This warning highlights the need for the paradigm shift discussed earlier.

Although disconnected in the early alternative food movements, the link between environmental and social justice has become sadly obvious and difficult to ignore again, even in developed economies (Lang et al., 2009). There is an increased awareness of how inequalities, climate change and lack of food affect the health of people on low incomes across the world, both in developing countries and also in affluent societies. The increase in food banks and food poverty in rich countries is also putting the social justice debate back on the policy agenda (Caraher and Dowler, 2014). Growing movements such as Via Campesina, with its principle of food sovereignty, are giving a voice to the social and environmental struggles of traditional and new peasants across the globe (Via Campesina, 2011).

The FAO has taken an active role in promoting a rights-based, participatory approach to fairer and sustainable global diets. In 2010, the FAO published the following definition:

> Sustainable diets are those diets with low environmental impacts which contribute to food and nutrition security and to healthy life for present and future generations. Sustainable diets are protective and respectful of biodiversity and ecosystems, culturally acceptable, accessible, economically fair and affordable; nutritionally adequate, safe and healthy; while optimising natural and human resources.
>
> (FAO, 2010a)

A year later, one of the most comprehensive attempts to outline the attributes of an SFS was produced in the format of a multiple values model that came out of the UK's Sustainable Development Commission (SDC) in 2011. The SDC, dissolved by the Conservative and Liberal Democrat coalition, had the role of advising government with regard to sustainable development and put forward a list of multiple values as part of its work in food systems. The list of values, shown in Table 3.2, was developed based on Lang's omnistandards (Lang, 2010; Mason and Lang, 2017) and includes six headings (quality, social values, environment, health, economy and governance), with several subheadings that act as a checklist for assessing the all-inclusive sustainability of food.

Efforts to define and quantify sustainable diets continue, both in the academic sphere (Garnett, 2016; Drewnowski, 2017; Mason and Lang, 2017) and in the NGO sector (WWF, 2017). Lifecycle analysis (LCA) of the environmental impact of foods is a common method for quantifying sustainability. LCA is based on the use of land, water and energy resources during food production and utilisation (Jones et al., 2016; Drewonoski, 2017). The shortcoming of an LCA approach is that it does not take into account the cultural factors that are intrinsic to diets (Sustainable Development Commission, 2011). Similar efforts to quantify the success of agricultural cooperatives by focusing on their financial performance, while ignoring cultural dimensions, are worth noting.

With regard to the link between sustainability and cooperatives, only one academic paper was found presenting evidence on how cooperatives are more sustainable than IOFs (Baranchenko and Oglethorpe, 2012). Other non-peer references reviewed included reports from the Scottish Agricultural Organisation Society and Co-operatives UK highlighting how cooperatives save resources by sharing equipment and how members are less likely to waste

Table 3.2 Multiple values for a sustainable food system

Quality	Social values
Taste	Pleasure
Seasonality	Identity
Cosmetic	Animal welfare
Fresh (where appropriate)	Equality and justice
Authenticity/provenance	Trust
	Choice
	Skills (for food citizenship)

Environment	Health
Climate change	Safety
Water	Nutrition
Land use	Equal access
Soil	Availability
Biodiversity	Social status/affordability
Waste reduction	Information and education

Economy	Governance
Food security and resilience	Science & tech evidence base
Affordability (price)	Transparency
Efficiency	Democratic accountability
True competition & fair returns	Ethical values
Jobs, skills & conditions	International aid & development
Fully internalised costs	

Source: Sustainable Development Commission (2011)

resources, as they are member-owners and waste has a direct impact on their finances (Co-operatives UK, 2012; SAOS, 2012 and 2017).

Despite the variety of models, lists and attributes found in the literature, there is a strong trend to integrate both the environmental and the food democracy agendas. Even though the concept of food democracy does not guarantee food sustainability, it can offer a guiding light to navigate through an arduous food paradigm transition.

A second finding was a gap in the literature with regard to the relationship between cooperatives (especially ACs) and sustainability. As this gap was identified in the food policy literature, this research set out to contribute to the debate by discussing the diversity of approaches to environmental sustainability and food democracy existing in the farming cooperative sector.

Summary

This chapter has presented the findings of a multidisciplinary literature review of ACs. Searches revealed that two disciplines dominate the academic study of ACs: history and economics. Economists mainly study ACs as an organisational model that *still needs to be improved* and is expected to evolve to become a hybrid of cooperatives and privately owned firms. The work of two key rural sociologists, Thomas Gray and Patrick Mooney, was identified as particularly relevant to this research, as it provides a less reductionist and more nuanced approach to the study of ACs that moves beyond a black and white portrait of these organisations and aims to unpack the tensions and contradictions they encounter. This finding highlights the need for a more comprehensive approach and a critical food systems lens for the study of ACs in order to build integration of the knowledge we have of them and the policies designed to regulate them.

Several gaps were identified in the food policy literature including the focus on consumers' cooperatives over agricultural ones, as well as the strong but limited number of studies on the increasing corporatisation of the sector and lack of research linking sustainability and ACs. The literature on defining SFSs revealed that this is an ongoing contested area of work. This book provides a contribution to this debate by offering an insight into the perspective of ACs on this topic, which is of vital importance given that ACs are powerful actors in European farming policy and key shapers of rural realities. How are ACs addressing – or not – unsustainable practices in their sector? ACs rely on farmers, and these in turn rely on natural resources; thus, one could predict a strong self-interest in addressing sustainability challenges. Following the introduction to ACs and review of the history and academic literature that has studied these organisations, the next chapter presents the theoretical and methodological approaches that informed the research presented in this book.

4 Why methods and theory matter when studying cooperativism and sustainability in food and farming

From critical approaches to crystallisation

A critical approach to epistemology

The research presented in this book started from an inductive position, seeking to build up an understanding of ACs from an alternative lens to the neo-liberal economic approaches that dominate the existing literature (see Chapter 3 for a detailed discussion of scholarly texts on the topic). Because of the limitations of existing theories of ACs to provide a framework for an integrated food policy analysis, the final theoretical framework was not fixed a priori but gradually emerged from document and interview data, and fieldwork. Nevertheless, following Miles and Huberman (1994), it was noted that a tentative, rudimentary framework was a good starting point to approach the field and to structure the rationale for the selection of case studies. A theoretical framework helps connect the researcher to existing knowledge, and vice versa. The review of relevant theory also offers a basis for a potential hypothesis and choice of research methods. This section describes the theories that informed the initial theoretical and conceptual frameworks. Because the final theoretical framework did not take shape until the data analysis stage, it is discussed in more detail in Chapter 10, as it emerged as an outcome of this research.

An interpretivist and constructivist lens for cooperativism and sustainability

Interpretivism and constructivism are related approaches that developed as a critique to the emergence of positivism in social sciences (Lincoln and Guba, 1985). Both interpretative and constructive approaches discard the existence of a "true" and mind-independent reality, assuming instead "multiple, apprehendable and equally valid realities" (Ponterotto, 2005:129). In the context of cooperatives and SFSs, cooperative members and other policy and civil society actors are active participants that interpret (and reinterpret), as well as construct (and reconstruct), the meaning of these realities and how constructs such as cooperation and competition are realised in everyday practices and power dynamics. A constructivist approach acknowledges the role that individual values and experiences play in the construction of knowledge, and, in the particular case of this research, the construction of definitions and imaginaries about what cooperatives and SFSs look like: whether they are defined in financial terms and also – or instead – with environmental and social dimensions. Following Berger and Luckerman, this research notes that "as individuals try to understand and interpret their situations, they construct meanings or decide what events mean, and then respond to these social constructions rather than to the world itself" (Berger and Luckerman, 1966:512). In this sense, by framing their actions (including choices) and perceptions of reality within their own individual values and experiences, food actors (farmer members in this case) are not only imagining but also creating and recreating ideas and embodiments of food sustainability. They do this through both their individual choices (e.g. farming techniques) and collectively, through their cooperatives' practices, adapting and shaping their reality to fit their own constructs that are in turn shaped and influenced by the financial and social constraints and context they exist in.

Additionally, this approach is particularly relevant for the study of the dual character of ACs: the organisational level (the cooperative) and the sublevel of individual members (the farmers). How are imaginaries about sustainability and cooperative principles constructed at each level? Do the imaginaries from each level overlap or clash? This constructivist approach has methodological implications, and adequate methods were selected to ensure that participants were able to describe their views of reality by sharing their own opinions and their own understanding of sustainability in their own words, rather than existing categories or descriptions being imposed on them (Baxter and Jack, 2008).

The underpinning theoretical assumption made when applying an interpretative and constructivist approach to the study of food policy is that all individuals – and ACs as representatives of individual farmers – are assumed to be active participants. Actors are perceived as having the potential to shape their systems of production and provision, as well as to influence other actors, while also often facing barriers and factors that shape how they behave. Actors in the food and farming cooperative sector and policy spaces include: farmers, cooperative members, processors, retailers, policymakers, consumers, scientists and civil society groups. The topic of study involves policy, but also strong human and power relations and dynamics and the knowledge and experiences of all participants, regardless of their affiliations or positions in organisational hierarchies. Rather than in individualistic isolation, participants – and their ACs in the case of farmers – were thought of as agents that interact in and understand the food system through constant transaction with their environments, creating theoretical spaces to consider how different actors construct their own meaning of cooperation and sustainability and their own version of what a successful AC and a SFS should look like.

Despite acknowledging the role of agency in shaping how cooperatives operate, actors are not independent to act as they wish at all times, but are embedded in interwoven environmental, production, processing, distribution, consumption and policy systems. Two concepts are key to those interactions: food democracy and policy integration. The first one, the concept of food democracy, was repopularised by Lang in the mid 1990s to refer to "the long process of striving for improvements in food for all not the few" (Lang, 2007:12; Lang et al., 2009; Lang and Heasman, 2015). Alongside food democracy, calls for policy integration have also been increasing, emphasising the need for food policies that take into account all stages of the food system, from production to consumption, in an ecological public health approach that links health and environmental concerns (Rayner and Lang, 2012). From the combination of the need for policy integration from an ecological public health perspective and the notion of food democracy, a new theoretical assumption emerges: any proposed "solutions" that help create a SFS that is able to provide healthy foods without damaging the environment must permeate all sectors and all actors for it to work; a prerequisite for this is that *all* actors must be able to access those *solutions*, regardless of their income or position/role in the food system, including consumers on low incomes and farm workers (Hinrichs, 2000; Caraher and Dowler, 2007; Holt-Giménez and Burkett 2011).

The history of public health and marketing has demonstrated that having knowledge of what the "right" food options are is not necessarily followed by changes in life and consumption patterns, as social determinants of health have a key impact on the number and type of healthy lifestyle choices available to people (Milio, 1976). These social determinants have deep implications for this research, especially when analysing alternative ACs promoting healthy and sustainable models, as will be discussed later. The case study methodology selected offered the potential to unpick how those determinants are affecting and shaping ACs, asking who is benefiting from AC policies and who is being left behind.

In terms of the analysis of policies relating to farming and specifically to agricultural cooperatives, this research follows Kuhnian thinking (1962) by assuming that policy analysis methods are never able to provide a totally neutral and objective understanding of issues, as they are limited by the theories and methods used to study them. Theoretical positions are cultural, historical and determined by the geopolitical context they emerge from, as data from the case studies will show.

The philosophical underpinning of the policy analysis approach in this research is also linked to the overall constructivist theorising that runs throughout the whole book, but attempts to avoid endless relativism (Berger and Luckerman, 1966). So, ontologically speaking, this research is located in the constructivist assumption that there are infinite realities out there, depending on who is looking when and from where; but these realities overlap, and, by collating data from diverse sources and informants, we can attempt to draw a silhouette of the overlapping areas and the marginal edges that are not shared. It is at these edges that "third spaces" emerge and create the right conditions for change and innovation to surface, both in policy and practice, as is the case of new governance models in multi-stakeholder cooperatives, which will be discussed in Chapters 8 and 9.

STEPS pathway influence

The STEPS pathway multi-methods approach, used to consider questions of sustainability, politics and development, inspired and informed this study. STEPS is the Centre for Social, Technological and Environmental Pathways to Sustainability, based at the University of Sussex in the UK. Following the same interpretivist/constructivist line discussed above, the STEPS pathway approach recognises that "who you are shapes how you 'frame' – or understand – a system" (STEPS, 2015). The following quote from STEPS describes the need to open up theoretical spaces to acknowledge and document those voices and initiatives that could be key to achieving more SFSs but, for several reasons, are ignored:

> Too often the narratives of powerful actors and institutions become the motorways channelling policy, governance and interventions, overrunning the valuable pathways responding to poorer people's own goals, knowledge and values. Our pathways approach pays attention to multiple pathways and, backed by a variety of practical methods, helps open up space for more plural and dynamic sustainabilities. It also aims to open up the political process of building pathways, which are currently hidden, obscured or oppressed.
>
> (STEPS, 2015)

The following characteristics are central to the STEPS pathways approach (STEPS, 2011):

- a systemic approach to issues, moving away from a reductionist focus that ignores wider interrelations and feedback loops among actors, practices and institutions;
- the recognition that there are many ways of "framing" – understanding and representing – a system, whether by international or national policy actors and networks, different advocacy and civil society groups, different researchers or local people; the pathways approach values all framings and strives to open up those that are hidden, obscured or suppressed;
- highlighting pathways to sustainability alternative to the dominant narratives that perpetuate social inequalities and environmental degradation.

This critical approach informed the methodology of the data presented in this book, based on the fact that this research aimed to analyse current thinking and tensions in ACs. This involves opening up theoretical and empirical spaces for those food producers who are trying different versions of cooperation and whose voices are often not heard because they do not fit in the current discourses of consolidation and competiveness dominating the AC sector. The selection of cases was done accordingly, in order to compare ACs at different points on this continuum, from ACs highly embedded in the industrialised food system at one end to those trying to disembed themselves from these relations by recreating new networks and ways to embody and reproduce their understanding of agrarian cooperativism. It was acknowledged at all points that this continuum is not linear in reality, and thus many ACs can fall in different categories (e.g. small but intensive members). The STEPS approach was also applied in the way interviewees were asked to define sustainability in their own terms, without pre-empting them with ready-made definitions. This allowed participants to articulate their own views, concerns and hopes for food sustainability in their own words.

Theoretical influences integrated in a complementary framework of analysis: Paradigms, New Peasantries and Open Coops

Based on the evidence relating to food sustainability discussed in Chapter 3, the main theoretical assumption underlying this research is that the global food system is in crisis and requires an urgent paradigm shift (Lang et al., 2009; Lang and Heasman, 2015). Owing to the breadth of the challenges facing the food system, an integrated policy approach is necessary to overcome the New Fundamentals discussed earlier (Lang et al., 2009). Patricia Allen has also called for a "unified vision" and a "whole system approach" to achieve an alternative and sustainable food system (Allen, 2004). For Allen, the key aspect is to reflect on "who and what are to be sustained and secured" (Allen, 2004:210). This sober reflection poses many questions: is it the poor who need to be sustained? Is it small farmers? Growth? Specific methods of farming? Three theories were selected because of their scope to reflect the political, social and multilevel governance dimensions of food policy in a complementary manner. This section provides further details on this complementarity.

Following the findings from the literature review, ACs can be mapped on a continuum that starts with small, niche models and ends with corporate-like ones (e.g. NGCs), and many hybrids in between. Based on this diversity, two initial complementary theoretical approaches were identified as suitable for framing this complex reality: food policy paradigms (Lang and Heasman, 2015) and New Peasantries (van der Ploeg, 2008). Both theories include elements of ecological public health (EPH) thinking, as discussed below. EPH conceptualises food in four dimensions of existence: the material (inputs, resources), the biophysiological (biological processes, from plants to human bodies), the cognitive (cultural messages and meanings) and the social (human interactions). These dimensions need to be balanced to achieve public health that is both beneficial for humans and the environment they are part of and depend on (Rayner and Lang, 2012).

In the context of addressing ecological health and the need for a more integrated food policy, Lang and colleagues refer to the fact that "consumers eat food as social, health and environmental acts simultaneously, whether consciously or not" (Lang et al., 2009:298). If this thinking is applied to agriculture, it could be said that farmers also grow food as social, health and environmental acts simultaneously, whether consciously or not. Next, these two theories are presented, followed by an introduction to the Open Cooperative framework, which was used to analyse the specific case of multi-stakeholder cooperatives.

Clashing paradigms

In *Food Wars, The Global Battle for Mouths, Minds and Markets*, Lang and Heasman depict the current food system as a battlefield where the dominant "productionist paradigm" (PP) – defined as the ongoing industrial model unable to meet current environmental, health and social needs – is being slowly transformed by two alternative paradigms: the "life-sciences paradigm" (LSP), in conflict with the "ecologically integrated paradigm" (EIP). Lang and Heasman defined the construct of Food Paradigm as a "set of shared understandings, common rules and ways of conceiving problems and solutions about food" (Lang and Heasman, 2015:17). This interpretivist definition is based on words such as "understandings" and "conceiving" merged with a constructivist approach (as discussed in the previous section) that emphasises the active role of actors interacting and sharing common rules and ways of thinking.

The authors use these opposing paradigms to describe two conflicting new conceptions of the food system and call for an urgent introduction of integrated policy (Lang and Heasman, 2015). On the one hand, the PP is still shaped by its historical roots in the Industrial Revolution, and its main focus is on quantity of production through the use of technology and intensive agricultural methods. At retail level, a key feature of this paradigm is market concentration at agribusiness, processing and retailing level. Health is considered consumers' responsibility, rather than being embedded in the food system.

Like the PP, the LSP builds on the commercial control of markets, commodities and also knowledge, through intellectual property agreements. It is a clear continuation of the PP's liberal market approach and dependability of external inputs, just using different methods ("same wine in a new bottle"), and does not solve any of the inequities of the current system. Scientific knowledge and a profit-making logic play a key role in the LSP, pursuing the commercialisation of science at all levels of the food system: for example, the introduction of genetic modification (GM) at production level and nutrigenomics as part of the latest developments in retailing and nutritional advice.

In contrast, the EIP is still defined as a niche paradigm, with less lobby power than the LSP. Criticised by LSP adherents for being backward-looking, EIP resists the introduction of certain biotechnologies such as GM (Lang and Heasman, 2015). This paradigm adheres to more holistic, non-intensive agricultural methods, rejecting monocultures. By promoting multifunctionality, the EIP accepts a wider role in preserving the biodiversity of both animals and plant species. In contrast to the PP and LSP, the EIP conceives health as embedded in the system, rather than as a commodity (Lang and Heasman, 2015).

The main concern of the paradigm theory is that the dominant policy paradigm inherited from the post-war period of the 1950s still retains its productionist character today and has become embedded in food policy institutions (Skogstad, 1998). The underlying logic of the PP was centred on subsidising scientific and technical research (which has become increasingly biosciences-based) to increase food production and shelf life and facilitate logistics and distribution at lower cost in order to increase profit (especially for retailers) but also affordability for consumers.

This PP, based on agricultural exceptionalism, made sense in an era of food scarcity, and it was incredibly successful at solving the Malthusian problem of feeding the rapidly increasing post-war population (Lang et al., 2009). However, having the single objective of increased production meant that this paradigm overlooked both the nutritional quality of the calories produced and the environmental damage associated with its industrial production methods. The policy emphasis remained inside the farm gate, and, in the meantime, consumers and choice became kings. Unfortunately, the limitations and negative effects of this paradigm are so obvious (see the discussion on the New Fundamentals above) that a shift is needed, and a new approach that is based on food democracy and able to secure food sustainability (social and environmental) must be adopted. This shift in thinking was incorporated into the methodological approach by working with a multidimensional understanding of sustainability and food systems that goes beyond reductionist metrics such as carbon footprints or yields. This theory also informed the selection and categorisation of ACs according to which of the food policy paradigms they were more aligned to.

New Peasantries against the Empire

The New Peasantries framework put forward by van der Ploeg offers a distinct, modern conceptualisation of the peasantry. This framework is relevant to the study of emerging agricultural cooperatives in alternative food system networks as their members often self-define as peasants, as data presented in Chapters 8 and 9 suggest. According to van der Ploeg, peasants have been theoretically ignored in favour of what the author calls the "Empire". This construct refers to a "complex, multilayered, expanding and increasingly monopolistic set of connections (i.e. a coercive network) that ties processes, places, people and products together in a specific way" (van der Ploeg, 2008:255). The Empire is thus used as a metaphor to refer to the power concentration in the global food system, ruthless profit-seeking and unsustainable practices. Resisting the Empire to different degrees is what characterises peasants, who, as van der Ploeg argues, are here to stay despite their neglect by theorists. The author conceptualises the *peasant condition* or *principle* as an approach to farming (*coproduction of food*) based on relations of cooperation and reciprocity, with control over the means of production. This control allows peasants to enjoy a relative degree of autonomy from global market forces (van der Ploeg, 2008).

The author recognises the heterogeneity of the farming sector and recognises "various degrees of peasantness" (van der Ploeg, 2008:36). This understanding of the current diversity of food producers and their conflicting interests and approaches provided a suitable structure to categorise the different types of AC explored in this research.

Van der Ploeg proposes three different types of farming – capitalist, entrepreneurial and peasant farming – depending on the degree of assimilation by the Empire. Those producers trying to reduce their reliance on the Empire are part of what in food policy literature are commonly known as alternative food networks. AFNs have been defined as "emerging networks of producers, consumers, and other actors that embody alternatives to the more standardised industrial mode of food supply" (Renting et al., 2003:394). The constructs of Empire and AFNs are used to present a set of findings from this research in Chapters 8 and 9 when discussing ACs that are operating outside global food supply chains and pursuing alternative indicators of success and performance.

Both the New Peasantries and the Food Paradigm frameworks acknowledge the contested terrain of food policy (including unregulated areas) and provide a theoretical space to debate the tensions in the AC sector identified during the literature review. By opening this space, the research process allowed a wide range of perspectives from different actors to be taken into account, in line with the active participant approach discussed earlier. Nevertheless, it was noted and acknowledged from the beginning of the research that neither of these two theoretical frameworks actually focuses on ACs. This limitation and how this research set out to overcome this theoretical gap are discussed in the next section and in Chapter 10.

P2P Foundation's Open Cooperative framework

Some of the case studies presented later are legally incorporated cooperatives breaking down the boundaries between growers, buyers and eaters. As opposed to conventional agricultural cooperatives with a single type of member (i.e. farmers), MSCs bring different groups of stakeholders, as distinct membership groups, under one shared cooperative enterprise to meet their diverse food-related needs. As mentioned above, it was identified that a more cooperative-specific theoretical framework was required to analyse the richness of these innovative governance models, as well as to be able to take into account these MSCs' links with other social movements beyond the cooperative movement.

The growing distance between producers and consumers, but also between consumers and the places where the food they consume is produced, has been long quoted as a serious concern in modern food systems (Kneafsey et al., 2008). As the literature review has revealed, the distance has also grown between the disciplines that focus on producers and farming and those that deal with consumers' issues. The study of multi-stakeholder food cooperatives presents an opportunity to bridge the common analytical gap that silos food production and consumption activities into two separate categories. With reference to this traditional theoretical gap, Goodman and DuPuis (2002) have called for an integrated analytical framework after reviewing the agri-food and food studies literature and realising the shortcomings of both production-centred perspectives and more "cultural" and consumption-centred theories that try to reclaim the consumer back into rural sociology. For the authors, "how the consumer goes about 'knowing' food is just as important as farmers' knowledge networks in the creation of an alternative food system" (Goodman and DuPuis, 2002:15).

This research attempts to bridge this gap while also acknowledging the specific case and context of consumer–producer relations in legally incorporated ACs rather than more informal associations (e.g. Food Assemblies[1]). In the modern industrial food system, food initiatives that foster producer–consumer interactions are normally labelled "alternative" (Holloway et al., 2007). Despite being a very helpful lens to look at farming and the wider food system, the two theoretical frameworks presented earlier do not offer enough specificity with regard to ACs in particular.

An additional gap identified was that the academic literature on MSCs is very scarce. This type of cooperative has hardly been theorised in the context of food and farming (Lund, 2012). Bauwens and Kostakis, from the P2P (peer to peer) Foundation, a forward-thinking international organisation focused on "studying, researching, documenting and promoting peer to peer practices in a very broad sense" (P2P Foundation, 2016), have put forward a fourfold proposal for Open Coops. This framework is general to cooperatives operating in different

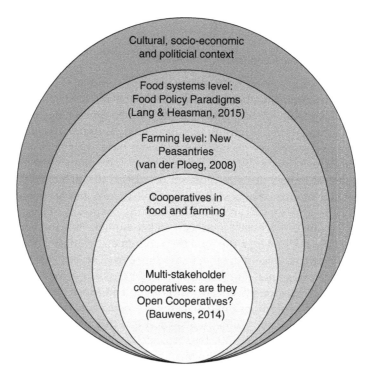

Figure 4.1 Multilevel theoretical framework
Source: Author

sectors of the globalised economy, with an emphasis on the information society (Bauwens and Kostakis, 2014). The open economy movement fights the increasing privatisation and commodification of knowledge, especially in the context of the internet age (P2P Foundation, 2016). Open economy activists are working to develop commons-based models for the governance and reproduction of abundant intellectual and immaterial resources (e.g. software and apps). At the same time, the P2P Foundation is working to link up with cooperatives as the ideal organisational type to develop a reciprocity-based model for the "scarce" material resources we use to reproduce material life (Bauwens and Kostakis, 2014). The vision is for the surplus value to be kept inside the commons sphere itself, creating a merger between the open peer production of commons and cooperative ways of producing value: "it is the cooperatives that would, through their cooperative accumulation, fund the production of immaterial Commons, because they would pay and reward the peer producers associated with them" (Bauwens and Kostakis, 2014:358). This research applies the Open Coops framework to the food and farming context. The Open Coop model is complementary to the theoretical frameworks discussed earlier (food policy paradigms and New Peasantries), as it adds a more refined dimension focusing specifically on cooperatives. Open Coops is a unique framework that calls for the evolution of the conventional cooperative model across four simultaneous dimensions:

1   Open Cooperatives should include in their own statutes their objectives and work aligned towards the common good, integrating externalities into their model.
2   All people affected by the activity should have a say (this is the specific multistakeholder nature of Open Cooperatives), practising economic democracy.

As Bauwens has pointed out, these two characteristics already exist in the solidarity cooperatives – another name for multi-stakeholder cooperatives (Lund, 2012) – such as the popular social care MSCs in Italy and Canada. The P2P Foundation framework advances two extra practices that MSCs have to incorporate in order to become meaningfully transformational (Bauwens and Kostakis, 2014):

3   The cooperative must co-produce commons for the common good, whether immaterial or material.
4   The final requirement is a global approach, to create counter-power for a global ethical economy consisting of cooperative alliances and a disposition to socialise its knowledge.

This framework is used to analyse MSCs in Chapter 8, in relation to case studies from both Spain and the UK. How do MSCs reconcile the different interests of different groups within an organisation? Can the MSCs studied be considered Open Cooperatives? Are they successful in serving the interests of the two weakest and most disconnected groups of actors in the food chain – that is, small producers and consumers? To what extent are they socialising their knowledge and joining forces to develop a global ethical economy and realise more-than-economic benefits of cooperation?

The theoretical benefit of combining the food policy paradigms, van der Ploeg's New Peasantries theories and the Open Coops framework is the complementarity they offer with regard to the levels of analysis: van der Ploeg's theory informed the organisational analytical level (the cooperative) to classify the different types of AC according to their food production strategy (level of embeddedness in the industrial food system). In conjunction, Lang and Heasman's food policy paradigms approach elevated the level analysis of ACs to the wider food system, providing a framework to assess their relationships with other actors (e.g. policymakers and consumers), as well as implications for sustainability. Finally, the Open Coops framework offered a specific lens to study the particular case of MSCs and their relationship with other social movements. These different levels of analysis informed the research process, in both the selection of case studies and the data analysis stage. This new approach to the study of ACs offered a powerful explanatory potential to analyse the political, socio-economic and environmental aspects of ACs, a much-needed contribution to the literature on the subject. The findings from the literature review informed the formulation of a conceptual framework that documents the key concepts and dimensions used in this research, as well as the relationship between tensions at the meso cooperative level and the macro food policy level discussed in Chapters 1–3 (summarised in Table 4.1).

Table 4.1 Tensions at the meso cooperative level and the macro food policy level

Dimensions of agricultural cooperation	*Related FP themes*
Legal: Informal co-operation–incorporation	Contested policy context Governance and MLG Notions of accountability Access to subsidies
Food system integration: Primary production–consumption	Danger of extreme localism Level of integration of actors National consumption vs. exports Links with consumers
Conception of food: Quality ("pride")–commodity	Quality vs. quantity Health implications Re-localism/environment Niche markets vs. food democracy
Reason for cooperation: Social motives–financial motives	Degree of food accessibility Ethics vs. practicalities Social capital and trust
Levels of identity/policy: Local–regional–autonomous communities/UK's devolved admins–national (Spain/UK)–EU–international (WTO)	

Source: Author

Seeking and acknowledging different perspectives: the role of triangulation and crystallisation

This section summarises the key aspects of the methodology employed to collate, triangulate and make sense of the data presented in this book. When two or more sources of data, theoretical frameworks and types of data collected converge and reach the same conclusion, that conclusion becomes more credible (Denzin and Lincoln, 1998). The concept of triangulation refers to researchers' attempts to use different avenues to explore if all sources of data and informants take them to the same final conclusions. This research project sought different levels of triangulation, including of:

- data: different data sources (policy documents, website content, e-newsletters, cooperatives reports), informants (cooperative members, policymakers, academics and civil society representatives) and collection methods (desktop research, fieldwork observation and interviews) were employed;
- theories: Open Coops were used to study MSCs; New Peasantries to frame the farming level of analysis; and food paradigms as a wider food policy framework;
- disciplinary approaches: the literature review analysed papers from food policy, economics, anthropology, sociology and biopsychology, as well as industry reports;
- cross-country: an extra layer of data triangulation was added by exploring the same research problem and questions at the European level as well as in two very different countries (i.e. UK and Spain), giving a cross-cultural dimension to this study, which also contributes to the quality, depth and richness of the findings.

Triangulation of data sources, data types and informants is a primary strategy that can be used to support the case study research principle that calls for phenomena to be viewed and explored from multiple perspectives. The inclusion of multiple, varied voices in qualitative research has been termed "multivocality" by Tracy (2010). This study has strived to follow a multivocal approach for several reasons: it is in line with the inclusive, critical STEPS pathways methodology discussed at the beginning of this chapter; it provides space for a variety of opinions and opens up space for viewpoints that diverge from those of the dominant majority or the author's; and it also suggests that authors are aware of cultural differences between themselves and participants, including differences in cultural background, nationality, class, gender and age, very relevant variables in farming, and so, credibility is enhanced when the research pays attention to these possibilities (Tracy, 2010).

Additionally, this research was informed by Cook's lists of techniques on critical methodological "*multiplism*" for capturing multiple views and sources of data, also in line with Tracy's multivocality principle: (1) embracing competing definitions of concepts; (2) using multiple research methods; (3) carrying out several studies of a single subject; (4) completing informal synthesis of loosely related studies; (5) using models with multiple causes;

(6) testing rival hypotheses; (7) involving a variety of stakeholders in the analysis; (8) adopting multiple theories or value frameworks; (9) promoting advocacy by multiple analysts; (10) and undertaking research with multiple targets (Cook, 1985:21–2).

The concept of triangulation assumes there is a single reality to be known and considers using a variety of data and methods as a way to uncover that reality. Several authors have highlighted the dangers of naïvely assuming that "there can be a single definitive account of the social world" and "*that sets of data deriving from different research methods can be unambiguously compared and regarded as equivalent in terms of their capacity to address a research question*" (Bryman, 2003:1142). However, triangulation was still considered a useful approach, as the advantages were expected to outnumber its disadvantages, and could enhance the credibility and validity of the research narrative and conclusions.

Like notions of reliability and validity, triangulation does not lay neatly over research from interpretive, critical or postmodern paradigms that view reality as multiple, fractured, contested or socially constructed. A more suitable term for research that uses multiple data sources, researchers and lenses based on post-structuralist assumptions is *crystallisation* (Ellingson, 2009). Richardson (2000) describes the crystal as a central imaginary that transcends the traditional notion of the rigid, fixed, two-dimensional triangle (Richardson, 2000). Crystallisation opens the door to a better understanding of a research problem from different perspectives, in line with the food systems thinking embedded in this study. In Tracy's words:

> Crystallisation encourages researchers to gather multiple types of data and employ various methods, multiple researchers, and numerous theoretical frameworks. However, it assumes that the goal of doing so is not to provide researchers with a more valid singular truth, but to open up a more complex, in-depth, but still thoroughly partial, understanding of the issue.
> (Tracy, 2010:844)

The aforementioned concept of multivocality is closely aligned with the notion of crystallisation, being an expression of it by including multiple, varied voices in the qualitative report and analysis. The multivocality principle informed the decisions about the selection of informants and how much data had to be collected, which in turn intersected with the level of analysis. The most important issue was to consider whether the data would provide for and substantiate meaningful and significant claims and perspectives. Multivocality emerged as an effort to analyse social reality – in this case, agricultural cooperatives' practices – from the participants' point of view (Tracy, 2010).

Transcription process and the use of quotes

A brief mention is provided in this section to inform the reader about "how to read" the quotes in this book. In a few instances, in order to add fluidity to the text, some of the quotes are presented as short extracts of interviews. In these cases, the researcher is referred to by an "R", and the participant is referred to by an "I" (interviewee). The reader might notice that a few quotes do not follow the standard word order expected in written sentences in English. This was done intentionally to respect and reflect the diversity and richness of interviewees' expressions and the individual nuances of their verbal communication, in line with the STEPS pathways methodology discussed earlier. Finally, square brackets in quotes ([. . .]) were used to provide additional clarifications to ensure the meaning of the quote was not lost or diluted in the translation and transcription processes.

Conclusion

This chapter has argued the need to reflect on the importance of theoretical and methodological approaches when attempting to study and "measure" complex social constructs such as cooperation and sustainability, to avoid reducing them in simplistic ways.

Acknowledging underlying epistemological and ontological approaches is key, as well as the need for critical methodological multiplism to capture multiple views from different groups of actors and sources of data. Special attention must be paid to including the perspectives of normally unheard voices, in this case, the accounts of small cooperatives that do not align with widespread indicators of success such as vertical growth and international activity.

The multilevel theoretical framework proposed to unravel the complexity of cooperatives in farming considers ACs as organisations embedded in the wider food and farming policy context, as well as being related to the global cooperative movement. Three theories for each of these perspectives of analysis respectively, include: New Peasantries, food paradigms and Open Coops.

Note

1 Food Assemblies allow consumers to place a weekly order directly with local producers, on a one-stop website. Food Assemblies are facilitated by a "host", who receives a small percentage of the sales of the assembly.

5 Experts' views on the European policy context

The price of remaining competitive and certifying sustainability

In 2011, the International Cooperative Alliance's list of the top 300 cooperatives revealed that many of the biggest cooperatives operate in the agricultural supply or food and drink sectors (ICA, 2011). Many were highly vertically-integrated businesses within the food chain. Several characteristics reflect signs of the increasingly market-oriented character of ACs in Europe, including control transfer from a board of directors to a management board (as the latter holds more market information), emphasis on marketing knowledge and innovation, review of financing structure to fund innovation and marketing strategies and, as a result, efforts to attract new investors with either voting (non/proportional) or non-voting powers (Bijman et al., 2010). This chapter discusses the price ACs are paying to remain competitive and sustain their market-oriented character. Their impacts affect not only ACs' members, but also sustainability in the food system. Next, attention is paid to the themes identified after analysing the data collected for this research: legal dimension, governance, sustainability, cooperative education, the importance of language, path dependency, social movements and future expectations for ACs.

Legal forms and cooperatives: a dilution of the cooperative spirit?

Before presenting the data related to this theme, it is worth revisiting here the general definition of cooperative (across economic sectors and activities) promoted by the ICA:

> An autonomous association of persons united voluntarily to meet their common economic, social, and cultural needs and aspirations through a jointly-owned and democratically-controlled enterprise.
>
> (ICA, 1995)

The key notions in the ICA's definition clash with the capitalist economies in which cooperatives exist, as, basically, a cooperative is "people-centred" rather than "capital-centred", the latter being the main feature of an IOF or private company (Birchall, 2010). In contrast with the ICA definition of a cooperative, ACs are defined by Co-operatives UK – the national trade body that campaigns for cooperation and aims to promote cooperative enterprises in the UK – in the following way:

> Enterprise co-operatives formed by individual commercial farmers working together in order to meet shared economic needs and achieve common economic goals.
>
> (Co-operatives UK, 2014a:2)

To complicate matters further, a relatively new form of farmers' association that has grown rapidly in the last few years is that of producer organisation. As discussed in Chapter 2, POs are economic organisations of agricultural producers (or fishermen) with characteristics similar to those of cooperatives (Bijman et al., 2012). EU legislation explicitly states that a PO can adopt any legal entity; a PO may have the legal form of a cooperative, but in many cases it has not, either because the legal requirements for cooperatives pose many restrictions on the activities and the structure of the PO, or, in the case of countries that have transformed from a socialist state economy, because the term cooperative has a negative connotation (Bijman et al., 2012). The next chapter will present interviewees' claims with regard to cases of abuse of the PO form in Spain. The following quote from a representative from the Plunkett Foundation reflects how the misuse of the PO model as a shortcut to access EU funding is in fact quite widespread:

> There are examples of people using it really well, but there are examples of people getting significant amounts of funding to be a PO and do what they do already, large businesses, you know, collaborating very informally.
>
> (Plunkett Foundation rep.)[1]

In this context, some authors have suggested that the historical incorporation of cooperatives into legal frameworks actually reduced their original transformative vision and has diluted the movement (Mulqueen, 2011). It could be argued that POs are both a result and cause of that effect. Guinnane and Martínez-Rodríguez have pointed out the underlying legal similarity of corporations and cooperatives, which "will surprise those accustomed to thinking of cooperatives as the very opposite of the corporation, and often formed to combat the power of corporations" (Guinnane and Martínez-Rodríguez, 2010:2).

The EU is pushing for the PO model over the cooperative model through its Common Organisation of Agricultural Markets' regulations. This preference for POs is not necessarily due to a specific disenchantment with the conventional cooperative model. A top agricultural policymaker from Spain suggested this is the approach taken by the EU because Brussels is not allowed to interfere in member countries' commercial laws. Brussels can and does push for and reward concentration of production, but it is out of its remit to impose the cooperative form to achieve that result. In short, it is allowed to use measures of support to encourage farmers to work together in order to concentrate supply in the face of increasingly organised demand from large retailers, but it is not allowed to favour a particular legal form – for example, cooperatives or IOFs.

The growing popularity of POs also raises issues for Cogeca. Cogeca is the umbrella group that represents approximately 40,000 agricultural cooperative organisations in the EU. Cogeca works in partnership with Copa, the network that brings together 60 EU farmers' organisations (Copa, 2016). Copa-Cogeca is known for defending the interests of larger corporate farmers and dominating EU advisory groups, including EU DG AGRI (Alter-EU, 2012; Corporate Europe, 2014). Cogeca supports the development of POs in general; however, as one of its top representatives told me, it cannot risk weakening the position of existing powerful cooperatives:

> If we set up a competing structure better than [Agricultural Cooperative x] the investment that this cooperative and the farmers have put into [AC x] would be endangered and we cannot accept this because we also represent [AC x].
>
> (Cogeca management rep.)

Cogeca sees successful ACs as those that offer the best economic returns for their members:

> Cooperatives [are] becoming stronger in the market place and becoming bigger; it is not necessarily 100% that if you become bigger you are stronger and vice versa, but it seems to be the tendency.
>
> (Cogeca management rep.)

For that reason, the sector and governments favour mergers and acquisitions. However, for campaigners fighting corporate control in the food system, the underlying ideology of the CAP is palpable:

> The CAP stands for Common Agricultural Policy. In reality, it is the Capitalist Agricultural Policy, and as such, it pushes what they want to push, they push a model of abandoning small farms in Europe and they open spaces for the agri-industry and to those models of agrarian colonisation, to those large companies, uniform consumption, etc. So those are the objectives of the CAP, and all the tools of the CAP, absolutely all, are addressed in that direction: the subsidy policies, the rural development policies, even the policies to foster new young entries into farming are designed to get young people to practice a capitalist agriculture [. . .] Agrarian cooperativism would work better without the CAP and without the European Union. Without "this" European Union.
>
> (*Soberania Alimentaria* rep.)

Other civil society representatives went even further, suggesting that the legal form of the cooperative is no longer transformative:

I think the legal model has failed. It does not respond to its original needs. And principally because the cooperative movement has not addressed the transformation of society and [cooperatives] have been islands in the framework of the capitalist economy, and in the end, in order to compete they have to enter in the same contradictions typical of private companies [. . .] and for their own members, the AC does not represent a big difference from the other agro-export companies.

(Via Campesina rep.)

This point about how large cooperatives' behaviour does not differ from private companies, and that many farmers feel they are detached from their own ACs and no longer have control over them, was also mentioned in the dairy sector report of the European Commission's Support for Farmer Cooperatives project that has been referenced throughout the text (Hanisch and Rommel, 2012). Furthermore, as the representative from the Plunkett Foundation pointed out, the CAP has fuelled polarisation in farming, rewarding both larger farms and larger ACs with fewer but larger members. Consolidation in this sense can be seen as a barrier to the regeneration of the sector:

it is difficult for new cooperatives to form because there is that massive incentive for POs of a certain scale than for a smaller organisation that is below that lower limit, trying to emerge and trying to be useful but they are at massive competitive disadvantage, so what we have seen is very few new cooperatives formed over that period.

(Plunkett Foundation rep.)

The pressure to become larger comes, not only from competition in the EU market, but also from the concentrated, powerful retail sector. Originally, farmers joined ACs to get more bargaining power when negotiating with buyers. Nowadays, the demand challenge posed by large buyers is not just about amount of supply and low prices, but also continuation of all-year-round supply, as this quote from a Scottish Agricultural Organisation Society (SAOS) representative illustrates:

We found that the concentration in the food processing sector has increased a great deal in the last few years, and that has meant that many of our coops have merged and consolidated into larger businesses so that they can supply food processors all year round.

(SAOS rep.)

The above point reflects how currently ACs fit the logistics model of large supermarkets well. By concentrating production and facilitating deals, it is much quicker and easier for a supermarket buyer to deal with one AC salesperson than with thousands of farmer members individually. ACs are also easy entry points to large groups of farmers. ACs can act as the trusted intermediate link to communicate with hundreds of farmers and encourage them to adopt specific standards to meet supermarkets' own line requirements. By inserting themselves into the market of a capitalist economy, cooperatives stop cooperating:

> If we take advantage of the legal form, the economic entity that says the cooperative is for profiting, then we enter another type of rationality which is not the cooperativist one, but a corporate rationality focused on economic benefits, mere profiteering, which is exactly what the cooperative model tries to avoid. A cooperative is not made for profiteering. A cooperative reproduces itself socially and economically. Then, when that rationality and therefore the ethics that underlay that rationality break, the essence of the existence of the cooperative breaks with it.
>
> (Spanish academic 1)

The data suggest ACs are increasingly opting for growth and investment strategies similar to those of privately owned companies (Bijman et al., 2012), at the expense of members, who lose control, and also of new farmers producing for the AC but who are not always offered the chance to become members of the AC they are contributing to (Berthelot, 2012).

Governance: the skimming of the cooperative principles

Cooperative principles are how cooperatives put their values into practice and what differentiate them from privately owned businesses. There are two main sets of cooperative principles relating to ACs. On the one hand, cooperatives in all sectors are expected to adhere to the seven internationally recognised International Cooperative Alliance principles discussed in Chapter 1 (ICA, 2015).

Chapter 1 also discussed how, during the 1980s and 1990s, the globalisation of the food trade exploded, fuelled by the inclusion of agriculture in the WTO in 1994–5. At that time, arguments for a simplification of the principles were put forward (Birchall, 2005) in order to "allow" ACs to be more competitive and attract large farmers. In 1987, the United States Department of Agriculture (USDA, 2002) adopted a simplified version of principles for cooperatives, specially designed for ACs (Ortmann and King, 2007):

1   The User-Owner Principle: Those who own and finance the cooperative are those who use the cooperative.
2   The User-Control Principle: Those who control the cooperative are those who use the cooperative.
3   The User-Benefits Principle: The cooperative's sole purpose is to provide and distribute benefits to its users on the basis of their use.

An aforementioned recent report from the USDA suggested the Department acknowledged how the USDA three principles were a "reduced form approach", developed through the "lens of economics" and prepared by economists to exclude values from the definition and identification of cooperatives (USDA, 2014:3). The paradox of removing values from the cooperative principles to emphasise their economic edge is that it dilutes the ethical and social essence intrinsic to cooperative identity. However, the repercussions of the diluted cooperative principles spread beyond the US. As discussed in the introduction, the user principles reached Europe, where they were promoted at the turn of the last century and are still used today (USDA, 2014); the last report commissioned by the EC in 2012 still defines cooperatives by listing the three principles proposed by the USDA in the 1980s (Bijman et al., 2012).

Birchall, a long-established author in the history of cooperation, has referred to globalisation in the food system as the main reason for the skimming of the cooperative principles in ACs:

> There is a view that those co-operatives that are operating in global markets – particularly agricultural marketing and processing co-operatives – cannot afford to internalise the ICA values and principles, but have to work with a slimmer and more self-centred set of principles just to survive; [referring to the USDA principles:] these are the kinds of principles that can be derived from the internal logic of a co-operative, and they have nothing to say about any wider social responsibility, but their market advantages may be clearer and easier to demonstrate.
>
> (Birchall, 2005:49)

This partiality for market advantages that are easier to measure and quantify will be discussed in detail in Chapter 10 as one of the biases that determine how the "success" of ACs is measured and how this is shaping the way the AC sector reproduces itself. Although some of the interviewees – especially from smaller and newer cooperatives – knew about the existence of the ICA principles and values, none of the cooperative members who participated in this research knew all the cooperative principles by heart. The fact is, several participants were not even expecting AC members to be aware of them:

> No, absolutely. And I would be surprised if they did. [. . .] I would be astonished if most farmers were aware of them individually, or that they would have learnt from them somewhere, or if they were on the wall of First Milk who are currently on the headlines, saying here are the principles; well, it isn't. They see them very much as cooperatives that give them some commercial advantage, some edge, some ability to withstand the shocks long term.
>
> (UK policymaker)

The short answer to that is that they are not aware of the principles and particularly, the sort of community and all the other stuff, because it's . . . what most farmers have now started to do essentially, is to work on the basis of they buy what is easily available and at the best price they can get it. And those two things apply. And I think that's where we are at.

> (Farmway member)

In their study of cooperatives in Scotland, Wilson and MacLean also found that the principles were not being adhered to in any strict sense (Wilson and MacLean, 2012). An ICAO representative explained to me his belief that this departure from the cooperative principles was increasing because "cooperatives in developed countries are much more focussed on business, and as result, they are losing some interest in governance" (ICAO rep.).

Evidence suggests that social capital decreases in large ACs, and members lose the sense of being owners of the cooperative, behaving more like customers instead (Nilsson et al., 2012). However, although all the farmers interviewed as part of this research were frustrated with the increasing corporatisation of their ACs, they also highlighted that they could not survive without them. A representative from the *Soberania Alimentaria* magazine in Spain put it in these words:

> It is true, even people very embedded in classic cooperativist models have the same critical reflection: [. . .] "it's gone out of hand", "we no longer have control over cooperatives". They don't know well why, but they criticise them because they're in the hands of directors; that cooperatives have grown too much; that cooperatives only have their eyes on large exporting targets; that cooperatives have been the entry door for the green revolution; that have been the entry door for GM.
>
> (*Soberania Alimentaria* rep.)

Sustainability: "Monsanto will swap you six for half a dozen"

It is worth repeating here the finding from the literature review discussed in Chapter 3 with regards to the lack of peer-reviewed evidence for cooperatives being more environmentally friendly than their private counterparts. Only one academic paper was found discussing their environmental advantages (Baranchenko and Oglethorpe, 2012). ICA and SAOS also claim that cooperatives offer more sustainable business models, not only socially, but also environmentally (ICA, 2011; SAOS, 2012).

Originally, when the European Common Market was formed, the main concern for policymakers revolved around modernising farming systems to increase yields. At the time, ACs took the initiative in technical research, transferring the risk from farmer to cooperative entities and supporting small farmers to meet EU requirements and be able to access new markets (Giagnocavo, 2012). Cooperatives are still key in introducing their numerous small-scale members to technical advances, such as new machinery, greenhouses and so on (Gómez López, 2004; Giagnocavo, 2012). This can be seen in a positive or negative light, as ACs act as entry points for different types of technology and inputs, including some controversial innovations such as GM or intensive growing methods. As we will discuss in Chapter 7, in Spain, Anecoop, the largest AC of citrus in the country, which exports around 80 per cent of its production to the EU, shares an experimental farm (the largest in Spain) with the University of Almeria, where they carry out trials with Syngenta and

Monsanto, two corporations globally infamous for their unethical practices and unsustainable products.

In fact, some interviewees argued that farmers in Almeria have lost sense of the land by growing food in greenhouses and on soil replacements. Almeria is a European leader in organic crop control (Giagnocavo, 2012). Although this seems like a positive development, an academic familiar with the region pointed out that it was, in fact, the health issues related to intensive industrial production under plastic that had pushed farmers to adopt organic methods, after realising the impact chemical pesticides were having on them and their workers. Whereas, for farmers, health is one of the triggers and rewards for adopting more environmentally friendly farming methods, for policymakers, sustainability is often translated in money terms:

> for us, the basis for calculating sustainability is the agrarian income. And one thing that has not been calculated is the agrarian income cooperativised or not.
> (Policymaker)

When this participant was asked which they believed to be more sustainable, the small or large ACs (see Figure 5.1), they answered:

> These ones [large ACs]. These ones [the small ACs] are not bothered. It is also true that they pollute less because their production is lower. The big ones spend a lot of money on ISO norms[2] and that kind of stuff. And the middle ones, what do they do? They either don't comply with the rule or spend money they don't have to try to demonstrate they are sustainable to be able to get closer to the big ones. Complicated. Cooperatives provide a lot of help, because in terms of economies of scale they provide a lot of help, they have a quality department that introduces [...] sustainability indicators . . .
> (Policymaker)

The above quote builds on the ongoing theme of middle-sized ACs, how they just do not fit in a market that either rewards competitive economies of scale or niche producers. Chapters 8 and 9 discuss in detail how small ACs are emerging in those niche markets. These small ACs are redefining food sustainability in terms of increased vertical autonomy for farmers, who at the same time are better connected horizontally with other like-minded actors in the food system (e.g. other growers, consumers and civil society organisations).

The next quote is from a member of a small MSC producing sheep milk products in the Basque Country. The quote refers to the previous ACs the farmer had been a member of and exposes the power ACs have over the growing methods of their members:

> [. . .] supply cooperatives whose businesses are linked to a livestock production model, and at the end of the day, they even promoted industrial farming because that was what they were making a living from, from selling concentrated feed.
> (Esnetik member 1)

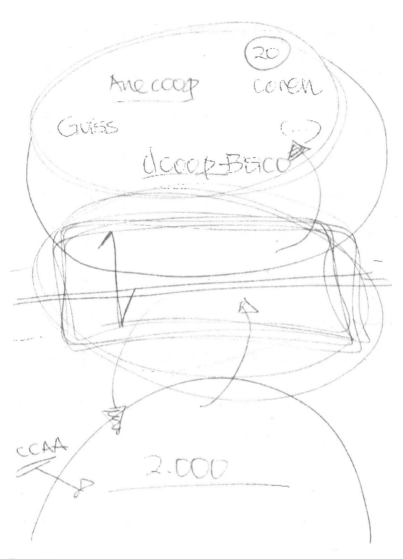

Figure 5.1 Drawing depicting the atomisation of the Spanish AC sector.
Source: Drawing by policymaker

Another participant also raised a similar point:

> There are a lot of big coops around the world that you would not hold
> up as examples of environmental sustainability, there are some that pro-
> vide multiple offers for their products in order to be able to sell farmers
> lots of fertilisers and financial products etc., but, I'm absolutely adamant

that coops help farmers to change, not all farmers want to change [. . .] but coops are a legitimate voice for farmers, so they are the voice of Openfield farmers but they are also a way of farmers telling each other what's coming and what's happening, so I think in terms of the environment, coops absolutely have a leadership role to say, ok, this is coming, there's massive instability in financial markets, there's massive instability in terms of weather patterns, in terms of commodity prices, all of this, all of this instability is a massive problem, are there ways of farmers doing things differently in order to help with that?

(Plunkett Foundation rep.)

Some ACs are indeed trying to introduce sustainability indicators. Going back to the middle-sized ACs and sustainability theme, the fact that the distance between consumers and producers is bigger than ever, and that many ACs are exporting their products, has created the demand for sustainability indicators that can be easily measured and demonstrated across distances:

Environmental sustainability is on the one hand what you are actually doing and on the other what you are able to demonstrate legally. We are getting to this stage. And if you are not economically sustainable, you are not going to be able to demonstrate your environmental sustainability.

(Policymaker)

Compulsory government policies to regulate growing methods to be more sustainable would benefit the environment but also large players:

We could introduce a compulsory life cycle analysis, and that will come at some point. What happens in Brussels is that they are very scared, like us. How many companies are able to maintain a sustainability study? Probably Mercadona[3] has the money to pay for this, and they will introduce measures and stop buying in Tunisia and will buy Italian or Spanish, but my problem is the same, SMES and middle-sized ACs, are they going to have money?

(Policymaker)

The lack of integrated policy affects ACs that find themselves walking a fine line trying to balance member benefits and surviving in the market. Those two objectives move social and environmental objectives down the agenda, as these quotes show:

You know basically we went back to the Co-op and told them that the possibility of a few farms being carbon neutral or negative or whatever, and their opinion was, there's a list of importance and at the moment it is down about 10th place, it is creeping up to about 7th, but can this have a premium, not really.

(Wales Rural Observatory, 2011:40)

> [Cooperatives] are not vehicles for furthering social and environmental change. They are very much vehicles for ensuring those independent farmers remain economically competitive and viable.
>
> (Co-operatives UK rep.)

For one of the former directors of Farmway, sustainability was a global, not a local, issue:

> The first question you've got to ask to sustainability is the sustainability of food production, so we have to know that the amount of food that is being produced and potentially, I mean I know there's huge wasted here, and potentially delivered to the consumer, is sufficient. And that means that has to be done in a world basis, it's not UK basis, world basis, because shipping food round the world is an equivalent of shipping water around the world, quite frankly [. . .] The second criteria to me is water, which is absolutely critical to the whole process, all over the world. And after that we get to what is generally termed environment, which is other species rather than man if you like [. . .] and the key question is that all over the world all farmers do exactly the same thing, and that thing is they harvest sunshine, that's what they do, and that sunshine that falls on that tarmac out there is wasted.
>
> (Farmway member)

Financial CAP subsidies can also be damaging for the environment, as is the case with olive production in Spain. This adaptable crop, which does not consume a lot of water, is perfect for the poor soils of non-irrigated areas in southern Spain (72 per cent of current production is still grown in these conditions). Unfortunately, EU subsidies have encouraged new, irrigated, intensive plantings, showing a lack of policy integration, as it ignores the environmental impact of irrigation in the increasingly dry southern regions where most of the production is concentrated (Giagnocavo and Vargas-Vasserot, 2012).

Cooperativas Agro-Alimentarias is the industry body that represents a large percentage of the 4,000 ACs operating in Spain. It offers training courses to farmers, and the quotes below refer to how sustainability issues are discussed (or not!) with members:

> Yes, they do mention that there are more pests because they are becoming resistant. Or if the farm next door has crops from a not trusted nursery, this can introduce pests if one plants things that are not certified.
>
> (Cooperativas Agro-alimentarias rep.)

When discussing the role of ACs in supporting their members to become more sustainable, the interviewee said:

> Well, the cooperative is involved with the members and the members are on the ground. I mean, what is needed is economic sustainability, because otherwise it loses its *raison d'être*, but also . . . the topic of phytosanitary

products or fertilisers, the member is going to be in that farm for 40 or 50 years, so one has to fertilise respecting the environment [. . .] they look at long term sustainability because the member is going to continue being there. There is a lot of work going on around managing integrated production;[4] that work in ACs does happen a lot, it is fostered a lot.

(Cooperativas Agro-alimentarias rep.)

However, Cooperativas Agro-alimentarias itself has stopped offering sustainability courses to its members:

Do you know what happens? For farmers, you organise a course on how to be more sustainable and people don't sign up, so those topics you pack them in . . . but in other different courses on offer, mainly around integrated production, but if you call it sustainability such as in "sustainable management", they don't come.

(Cooperativas Agro-alimentarias rep.)

Could this be farmers' rejection or distrust of the word sustainability?

That is why I'm telling you that this topic of sustainability or sustainable management is not a word that fits with them well. [. . .] The cooperative has to foster the sustainability of members but without organising courses on sustainability, they have to tell them about it through training, but the course is not called like that, but something along the lines of "fertiliser of whatever" and then you tell them.

(Cooperativas Agro-alimentarias rep.)

The interviews revealed that sustainability is a concept many farmers are suspicious of. Some of the farmers I spoke to believed the word is used to justify and impose measures to further strengthen the industrialisation of food cultivation and to be sold new products, or in a farmer interviewee's own words: "Monsanto will swap you six for half a dozen".

On the other hand, interest in sustainability can act as a point of convergence for farmers who are currently locked in by their own ACs' practices but who share an interest in exploring new forms of production. This research found that organic farming can actually bring farmers from different ideologies together. For example, only 1 per cent of cherry farmers in the Jerte Valley in Spain (where approximately 80 per cent of the national production of cherries is produced) farm organically. Of those, some converted as a way of opposing the general political and economic system. Others, through collaborations with trained agricultural agronomists, came to understand the negative impact their production methods were having on the valley and on their and consumers' health. A few of those farmers have now formed a small AC to commercialise their organic production.

At the EU level, Copa-Cogeca often discusses sustainability, but from a more reductionist perspective. In a presentation covering the challenges stopping ACs from adopting more sustainable practices, Copa-Cogeca high-lighted the market not remunerating the provision of public goods as the main barrier. The umbrella organisation suggested that financial incentives within and beyond the CAP are needed, as the CAP is neither a research nor a climate change policy (Copa-Cogeca, 2011). As part of the solution, they called for a focus on sustainable intensification and private–public investments (Copa-Cogeca, 2011).

There are many other examples of Copa-Cogeca's preference for sustainable intensification strategies. In March 2015, Copa-Cogeca, along with 13 other trade organisations trading animal feed, published a review of the EU's deci-sion-making process to authorise GMOs for food and feed uses. In this posi-tion paper, the trade network, claiming to represent "the whole of EU food and feed chain", urged the European Commission to accelerate approvals of GMOs for animal feed (COCERAL, 2015a:1). Copa-Cogeca has also pushed the EU to allow GM import authorisations. The EU Food and Feed Chain partners (which include Copa-Cogeca) rejected an EU Commission proposal that attempted to renationalise EU market authorisations of GM crops for feed and food use (COCERAL, 2015b). Copa-Cogeca's view was that it threatened the internal market for agri-food products. A month later, Copa-Cogeca's net-work labelled the EU's zero-tolerance policy as a "threat to competitiveness".

In 2008, at the EU Joint Research Centre's Workshop on the Global Commercial Pipeline of New GM Crops in Seville, Cogeca called for tem-porary approvals for all GM livestock products risk-assessed by the European Food Safety Authority and for those having being authorised based on Codex guidelines. Failing to do so would have a negative impact on European farmers, as they would lose competitiveness, Cogeca claimed (EU Joint Research Centre, 2008). With regard to the recent Transatlantic Trade and Investment Partnership (TTIP) negotiations, Copa-Cogeca, while seeming to be very supportive of the TTIP negotiations, was worried about the different levels of standards. In a joint document from Food Drink Europe and Copa-Cogeca, the umbrella organisa-tions raised their concerns about European citizens expecting farmers to imple-ment increasingly expensive and higher production standards, while imported products would not have to meet the same requirements. They believed this inconsistency had to be resolved in the TTIP agreement and, as their joint posi-tion document stated, their ultimate goal was avoiding any new regulatory bar-riers for EU farmers being introduced (Food Drink Europe and Copa-Cogeca, 2014).

In summary, this section on sustainability has shown how, on the one hand, ACs can offer continuity and an ability to help members think long term. As short-term policies are fuelling environmental degradation, long-term approaches are much needed. On the other hand, many ACs seem to have become active agents and victims of short-term pressure and are now pushing their farmers to produce faster and more intensively. This is especially so in the case of supply cooperatives making a living from selling pesticides and fertilisers to members.

A quote from the ILO will be used to close this part of the discussion on sustainability. At the end of 2012, the ILO updated its guidelines for cooperative legislation. The book, in its third edition, was authored by Hagen Henrÿ and is based upon the ILO's Recommendation 193 on the Promotion of Cooperatives. The book's motto, "Cooperative enterprises build a better world, but cooperatives cannot – and must not - save the world", emphasises that cooperatives should not be seen simply as a panacea in crisis situations (Henrÿ, 2012). This quote and the data presented in this section raise several questions: How can cooperatives provide a balance between short-term and long-term objectives? Are the cooperative legal and governance models able to help or balance the human and policy preference for short-term rewards? Or are we expecting too much of cooperatives by suggesting they should encourage members to adopt more environmentally friendly, fairer farming? Are cooperatives just suffering from policy stretching, being expected to do too many things, and save both members and the world?

Cooperative education

During the twentieth century, the professionalisation of farming and, thus, of ACs became a global phenomenon (Planells Orti and Mir Piqueras, 2004). In Europe, multilevel food policies are rewarding the consolidation of cooperatives (and POs), creating fewer but larger players (Bijman et al., 2012). Smaller producers wanting to practise a more ecological and socially just way of farming are having to reinvent the cooperative model or/and seek help and funding from civil society organisations (see Chapter 9 for a detailed analysis of this trend). As an academic expert pointed out during an interview:

> It is a shame that professionalisation is confused with capitalisation. [ACs] do not become more professional by becoming larger or by having a more . . . *hipster* logo.
>
> (Academic 1)

Farmers are constantly receiving a contradictory message from industry and multilevel governance policies encouraging them to be both competitive and cooperative at the same time. Being competitive in the food system has become a virtue in the sector but has raised contradictions in ACs:

> then there were periods when the [UK] marketing boards were thought to be anti-competitive, a really decisive point was when the marketing board was divided into different cooperatives. The emphasis was on the cooperatives should compete with each other, but when you look at the cooperative principles, they state that cooperatives should cooperate with each other.
>
> (Plunkett Foundation rep.)

Some participants highlighted how, in most ACs, there is a disregard for keeping the cooperative vision alive among members, both by making them aware of the origins of their own ACs and the wider vision of the cooperative movement:

> The most common issues that seem to come up is not what they are saying, it is that they do not really understand what it means to be a cooperative member [. . .] our founder always said that there was cooperation and education. Cooperation was the doing and setting stuff up. Education was the ongoing challenge of getting people to understand of what he called the profound implications of cooperation. So it is an ongoing challenge [. . .] educating [the] membership to understand what it means to be a cooperative [. . .] to get people to understand what it means to be a member.
>
> (Plunkett Foundation rep.)

Education and training for members, not on growing methods or business techniques, but related to the cooperative movement, were considered key by several interviewees:

> If cooperativism is a way of struggle, of work, of securing a job, if we don't maintain it and the cooperatives themselves don't invest in training that explains the reasons of cooperativism in an ongoing way, our children are going to lapidate cooperatives because they have not felt the want and they see it as an industry. Training is a basic, and in order to deliver good training, it needs to be budgeted and that is what is happening to the Mondragon group, the lack of ideology, the why of cooperatives.
>
> R: but they have a university?
> I: yes, but they don't cover these topics, it's more the process of production than the ideology.
>
> (Cooperative member and Via Campesina rep.)

The problem of short-term cooperative memory worsens with generations. A rural sociologist interviewed as an expert informant put this phenomenon in academic terms:

> The issue is that if you do not participate in the origin, the birth of an initiative, and especially when it is born from resistance, and that oppression is experienced and that gives way to the social creativity that allows that initiative to emerge, and in this case of cooperative x, of course, you inherit all that ready-made work.
>
> (Academic 1)

The relationship between this reductionist understanding of professionalisation and the lack of cooperative training is well reflected in this quote:

> It is very easy for a cooperative, as it becomes more complex in its operations, for the technical and managerial staff to start dominating. And not to invest in reproducing the solidaristic elements from which often coops have emerged. And the question is, how long it takes for that development to happen. I think the reproduction of the spirit of cooperation is the hardest challenge in any of the coops I have ever worked.
>
> (Academic 2)

With remaining competitive being the main aim for ACs, cooperation becomes a specific means to an end, seeking continued short-term profitability as opposed to cooperation as part of a social movement, with the final ambition of transforming society into a cooperative one (Oakeshott, 1978; Mooney and Majka, 1995). However, imposing more formalised governance for ACs that regulates and monitors their adherence to all the cooperative principles was not deemed an easy or advisable solution by participants:

> I would completely avoid the legal obligation side because cooperatives are their best when people agree voluntarily to do things, people come together voluntarily, work together voluntarily, there is no force, no one is forcing them to do things, because otherwise you end up being like a collective farm, end up being like a Russian collective farm which is not a cooperative at all, so the informal education side is what we focused on rather than the formal *"you must tick this other box"* because most people don't ever read that anyway.
>
> (Plunkett Foundation rep)

> I think ACs are often being outside the movement for that reason really, it's because people have said, well you do not have set social objectives, how do you demonstrate concern for community and all of that and it has meant that the movement, that part of the movement has been isolated.
>
> (Plunkett Foundation rep)

The importance of language and terminology

Closely related to the previous theme but distinct enough to deserve a separate heading is the theme of language. The importance of language in transforming the identity of ACs was also identified as a common and important topic in two country cases covered in the next two chapters. In Spain, the 2013 law designed to foster mergers and acquisitions in the AC sector formally introduced a name change to refer to ACs, swapping the previous more sociological term "agrarian cooperatives" for the more enterprise-like "agri-food cooperatives". This was the reasoning behind the change:

[To] improve the definition of the types of agrarian cooperatives in their current name to adapt them to their economic and social reality, from now on, they will be referred to as "agrifood". [. . .] The new name agrifood cooperative that is given to agrarian cooperatives has been called for by the totality of the sector and it is adequate given its closer approximation to reflect the socio-economic reality. This modification is extended to every instance of the name agrarian cooperative in the Law 27/1999 of 16th July.

(BOE[5], 2013)

Disregard for cooperative principles becomes a normalised trend when the bottom line is presented as the main priority. For certain participants, language was considered to be very important, and some thought farmer-owned businesses should only be allowed to call themselves cooperatives if they complied with the ICA principles:

> If there were truth in the terminology, many cooperatives would be sanctioned for using that name. And it would be right that they were sanctioned or that the use of the word was limited and they had to present themselves as they are. Hojiblanca cannot say is a cooperative and they use the name, they use it because from a commercial or marketing point, is in their benefit. It would be good, without getting obsessed, to clarify the use or delimit the use, to each their own.
>
> (*Soberania Alimentaria* rep.)

In the UK, two members of smaller farming cooperatives aligned to ecological farming and food sovereignty principles agreed that large ACs, distant from their members and embedded in industrial and global food conglomerates, should not be allowed to call themselves cooperatives. However, they could not suggest if or by whom and how the use of the word "cooperative" should be policed. Another interviewee offered a potential solution:

> at least we can start using it, ourselves, for example, we talk about industrial agriculture, so I can for example try to talk about the "non-agriculture" or the "capitalist agriculture", but to be clear that that is not agriculture, that is something else. Then, in the same way that we can incorporate a language with a gender perspective amongst ourselves, let's incorporate statute language as well. Whenever we talk about Hojiblanca, let's talk about the "non-cooperative Hojiblanca", or I don't know, at least to be aware, right? [. . .] It is not urgent, as I said before, let's not get obsessed about this, but every little helps.
>
> (Soberania Alimentaria rep.)

For other participants involved in policy, the enterprise aspect of cooperatives and their financial sustainability, rather than their names or the language used, are what really matter:

those people, regardless of whether they are treated well or badly, if their cooperative is not financially sustainable, the rest doesn't matter. A cooperative enterprise is a cooperative but it is an enterprise! Especially with what's going on.[6] Of course the principles are very nice or very utopic and ideal too, I wish they were complied with, but the thing is, if they don't get moving, at the end of the day it does not matter what name they call their cooperative society, they are going to have to close.

(Spanish Policymaker)

This topic brought back to the conversation the strong political connotations and legacy that ACs still have in Spain:

This aspect that you are talking about is very nice, it is true that it's cool they call you by your real name. That yoke no longer exists, luckily or unluckily this is a democracy and we no longer have that kind of problem,[7] we have worse problems, in Spain for example is the politicisation of cooperatives at the municipal level that is Dantesque. Small cooperatives are the bone of contention of the town mayors, and that is shameful. [. . .] The cooperative leaning to the right supports PP and the lefties support PSOE, that can't be right! Then they give subsidies to the left-leaning one but not to the right-leaning one and the Diputacion does the same thing and the Autonomous Community does the same thing. So we have to open people's minds, that managers comply with the seven principles but that they must comply with the law first. At the end of the day is a matter of financial, environmental and social sustainability. Besides, there is a big problem of politicisation in the boardrooms of cooperatives, and in the small ones, it is worse, even worse.

(Policymaker)

This account reflects how historical events create certain paths of development that reveal the legacy of political conflict. This conflict is still very present today in the way ACs operate, at least in Spain, the way they are funded and the way they resist the ongoing top–down push for consolidation as ACs of farmers of opposite political views refuse to merge. This invisible path and how it shapes the development of the AC sector are discussed in more detail in the next section.

Path dependency and inevitability of the corporatisation of the AC sector

Building on the discussion from the previous sections on sustainability and the impact of history in the development of ACs, this section presents arguments made by interviewees with regard to the inevitability often felt by many participants about the way things are currently shaping for ACs, and how radical change is unlikely unless a crisis or breaking point arises. The concept of path dependency is used to discuss interviewees' comments on this theme.

Path dependency is a concept that combines historically informed scholarship and political science approaches (Kay, 2003). Kay has used this term to analyse the development of the CAP and defines path-dependent systems as those in which "initial moves in one direction elicit further moves in that same direction" (Kay, 2003:406). In the context of this research, the concept will be used to discuss the themes emerging from the interview data relating to how specific events – such as the incorporation of the UK and Spain into the EU, the incorporation of agriculture into the WTO, or the creation of certain cooperative laws – create a narrowing down of policy options in the future.

For example, some participants considered EU policy on POs a key factor affecting the AC sector and discouraging farmers from setting up smaller cooperatives. Others, however, thought the cooperative ideal was just utopian, while a different view was that ACs are not able to meet the AC principles because of the current dominant model of industrial agriculture. Also linked to this concept of path dependency were several comments around the unavoidable corporatisation of the AC sector; this quote from the Spanish policymaker interviewed is an example of the latter:

> This is not just happening because of the EU, this is globalisation, modernisation. The utopic ideal of a cooperative is very good, but at the end of the day, they are NIFs[8] understood as enterprises, and people come to them to make money [. . .] At the end of the day the aim is that the product, the goods and the service sold by a cooperative are slightly better in price to those sold by a guy not from a cooperative. And that is my objective. [. . .] At the end of the day I'm talking about money, how? Economies of scale, which cooperatives are very good at. [. . .] Moreover, what cooperatives do is to diversify the risk, because it is shared across all the cooperative members, but they have to make money. Cooperatives understood utopically like those invented in England, well, they're very good, but if you don't make cash, they're good for nothing.
>
> (Policymaker)

> Furthermore, in Spain the myth has died, because since Fagor[9] collapsed, well, I was very depressed, the example of social economy in all of Spain and Europe, the biggest Spanish cooperative and they go to hell for not doing it properly.
>
> (Policymaker)

When I highlighted that actually around half of Mondragon's enterprises are not employee-owned, the participant pointed out that this is a common practice among other large ACs too, such as Decoop and Anecoop (the case study presented in Chapter 7):

> That is the same with Decoop; Decoop, how many brands does it have? Anecoop also buys produce from non-cooperative enterprises.
>
> (Policymaker)

As the Spanish policymaker mentioned concentration of agricultural production several times during the interview, they were asked if they thought that was the general objective of the cooperative sector, to which they replied:

> No, the objective is to improve the agrarian income. So that this guy is able to have a better life and that he stays in the countryside producing agricultural goods and services, through what? Improving his agricultural income. At the end of the day it is going to be a macroeconomic datum. [. . .] It is basically a policy of good government towards your constituents but also selfishly because these guys give us the agrarian goods that foods are.
>
> (Policymaker)

Other interviewees highlighted the impact that path-dependency effects were having on farmers' lives. As discussed in the previous section, one of the interviewees reflected how the adoption of more ecological methods (not always certified) in Almeria was pretty much the only solution to improve working conditions and reduced the serious effects of pesticide exposure under plastic that farmers were suffering from and had power over. However, the organic farming they converted to was not the transformative one envisioned by the early organic movement, but one highly embedded in international supply chains, involving intensive production, several harvests a year and monocultures (Fromartz, 2007; Goodman et al., 2011). More radical changes in their way of farming and selling their produce would be much more complex to implement, and thus the current greenhouses and cooperatives' contracts keep them locked into, perpetuating their embeddedness in industrial and globalised food systems:

> It is the way chosen to survive, economically embedded in the economic model of exports, that is why they are trapped, there is not an easy political exit against this option, it is not a technical question, it is not a technical question. If you design your farm, your life, your family, your mind for export activities, something has to happen to make you change, something serious, something big that makes you move to get out of there, if not, you don't come out, because you're designed for that, not just your farm.
>
> (Spanish academic)

Several AC members spoke with sadness about the locked-in situation many farmers find themselves in. Some reported feeling trapped and unhappy with the way their ACs were running but acknowledged their farming livelihoods could not survive without them. These feelings of inevitability spread beyond sustainability topics into issues of scale and barriers to creating alliances between farmers and consumers, as the following quote from an SAOS representative shows; when asked about the potential of MSCs that bring together farmer cooperatives and consumer groups into one organisation, the interviewee replied:

I think part of the reason as well is that many farms in the UK in the last 50 years just got bigger and bigger, larger and larger farms and they have massive volumes of products to sell, so they have to be linked into supply chains that can take all that volume. So very often, they are linked in to supply chains that will be major food manufactures and major supermarkets so that they know they will be able to sell all their produce, so it's scale, there is a mismatch of scale quite often between farmers, coops and local food requirements. The two scales are very, very different. And maybe over time, we will see a trend emerge of I think more local multistakeholder cooperatives between farmers and consumers and restaurants, but it will take some time to come. Maybe the closest we've got to that is farmers markets where we have farmers bringing their produce directly to consumers.

(SAOS rep.)

These issues of scale also inform the types of member that large ACs look for and accept on their books. For example, some ACs specify a minimum amount of trade that members must do with the AC in order to qualify for membership. A small or medium producer with polycultures of different crops or types of produce is unlikely to reach the minimum amount of trade required by large ACs. How does this fit with the ICA open membership cooperative principle?

You are giving me a real work out, aren't you? [laughs] How does that fit with the open membership principle? Ah . . . it doesn't . . . ah . . . often . . . I mean, the limit around . . . you must do a certain amount of trade, I think it is an issue, cooperatives do it for logical reasons, they need a certain amount stuff to sell, and they can't get it from too many places, otherwise that can be difficult. So people below that level often have to look at other options, other cooperatives or if there aren't other cooperatives that willing to do that, they have to go Hello?? Which I don't like. I think smaller farmers need to have a place in cooperatives. If that isn't there, then there's a need to be different, different cooperatives need to be set up. And then we come back to the EU influence . . . this isn't across all sectors, but particularly in horticulture, there are big disincentives to set up smaller scale stuff because of the PO organisations bias.

(Plunkett Foundation rep.)

Here, POs were again mentioned, along with large, powerful retailers, as the culprits shaping a policy and trade environment that just pushes ACs to get bigger or get out. The following short extract from the interview with the policymaker also reflects an element of resignation and inevitability:

R: Via Campesina, do you know them?
I: Yes

R: Do you work with them?

I: No. You need all sorts of people . . . I go back to the same thing, utopia is necessary, we need to pursue some objective, but [. . .] I'd like to see them here. Like Podemos. They are logical sociological options taken into account what's going on. Cooperativising reality would be wonderful . . . we are giving a step to cooperativise part of the agrarian expertise [laughs].

(Spanish Policymaker)

Another participant who had been a member of a workers' cooperative shop before joining a farming cooperative shared their views on why workers' cooperatives are not popular in agriculture:

I don't know. Well, I have thought about it. That's why I started saying I don't think we've been reaping the benefits. I think the way that farming works is quite hard to act as a cooperative because [pause] I don't know, I haven't quite figured it out in my head, I think farming lends itself perhaps a bit more to partnership or sole trade kind of businesses.

(Worker cooperative food grower)

When asked if the structure of landownership was a factor, as well as farms being in the family for generations, the participant said:

Yes, there's that. But it's also the nature of the work. It's quite an isolated job [. . .] I think it's quite easy for most people who work on the land to just want to do things their way and to hire other people to do it. I haven't worked it out. Maybe it's the type of people doing it, rural . . . ? There's a certain stereotype of what it is to be a farmer and to have a farmer mentality, perhaps more closed-minded.

(Worker cooperative food grower)

Most of the food growers interviewed that were members of an agricultural workers' cooperative had been previously members of a workers' cooperative in other sectors (mainly retail). This was identified as an interesting finding. It suggests that previous experience in cooperative working is what other farmers are lacking. By working in a cooperative on a daily basis, alongside other members, individuals embody cooperation and some then take it into farming. However, many farmers and agricultural workers have not experienced such a way of cooperating and fail to see the value or the potential of workers' cooperatives in farming to reduce entry costs into this sector. This view is reinforced by the landownership structure and housing prices in rural areas, which means many farmers live on the land they farm, rather than travelling to a farm to work in the same way office workers travel to their offices.

Another interesting point of discussion with participants was around the feasibility of including a representative from consumer organisations on the board of directors of ACs. This strategy could avoid a polarisation of the sector or/ and ACs' growing focus on exporting produce. However, this addition would be highly complex in legal terms. In Spain, for example, the law already determines very specific roles for cooperative boards. Although some of the more progressive AC members thought this would be a good idea, the policymaker highlighted the legal obstacles to such an initiative:

> it is not feasible. Would it be a good idea? Totally, especially because it is an external opinion [. . .] but it is not feasible and it is not going to happen, moreover, it is not legal.
>
> (Spanish Policymaker)

In this context, the legal path of cooperatives is preventing potential innovative solutions that could help ACs consider more sustainable practices. Some participants felt the problem was not actually ACs becoming more corporate or the dominant model of intensive agriculture, but capitalism itself:

> They can only explain themselves or can only understand themselves as exporters, because it is the only way of surviving, it is not an answer specific to farmers of the farming sector. It is the general answer of a capital economy and they are in the frame of, we all are in the frame of that capitalist economy, I think this is an important point in order to understand this. It is the same answer that the PP or PSOE or any other political party to getting out of the crisis: produce more, internationalisation, improve competitiveness . . . Then, with those clues as an umbrella, the farm, whether it is a new farmer or a long-life farmer, if that is the environment, if that is what is being heard, if that is what 99 per cent of the politicians repeat, from the political administrations, from the managers, from business men, everybody only knows how to talk about growth and continue to progress, so if this is the general answer, they replicate the same answers.
>
> (Spanish academic)

> I think members of each cooperative should reflect on how they have arrived to the situation in which they find themselves now, but I don't think we need to do a blaming-placing analysis of cooperativism in itself; if cooperativism is what it is now, it's because society has taken it down this route, because they teach what they teach in schools and they teach what they teach in universities [. . .] we have delegated many things and this is what we get now.
>
> (*Soberania Alimentaria* rep.)

> Well, that's the problem of living in a market economy. We can't escape it.
>
> (SAOS rep.)

It is interesting how interviewees' answers to questions that had a specific focus on ACs ended up referring to wider structural issues, politics, capitalism and how dominant ways of framing ACs become perpetuated by the education system.

Using the concept of path dependency, this section has discussed how the current agricultural policy and economic context in which ACs exist have had a strong influence on the shaping of farmers' cooperatives. When certain laws, policies and practices are adopted, other alternatives are rejected or made more difficult to adopt in the future. The return to those other options is not always easy (e.g. converting from conventional to organic agriculture or from outdoor to greenhouse farming). The narrowing down of potential options precludes other alternatives from happening, as going down certain policy routes might require undoing many other decisions and polices with wider implications. Changes to the system (policy, social or economic) remain within controlled and specific boundaries, creating lock-in effects (Wilson, 2008) and hindering system transformation, one of the original aspirations of the cooperative movement, a theme discussed in the next section.

Social movement

This research focuses on ACs within the European context, but, as stated in the introduction, it aims to extend the analysis beyond their organisational limits and into the wider context of global food systems and the international cooperative movement. Despite being based in Europe, many European ACs' operations and practices have effects internationally (Bijman et al., 2012). In this sense, some authors have argued that the legal incorporation of cooperatives decreased their political engagement and their identity as a social movement (Mulqueen, 2011), a trend that has been aggravated in ACs by the neoclassical economics discourse (Mooney et al., 1996). The effect has been noted by the wider cooperative movement:

> In developed countries, ACs are very focussed on their business. I know they used to be very strong members of the ICA, but the ICA could not provide any profit for their business, so they withdrew all their members, so we don't have many ACs in Europe. The cooperative movement in general is a movement, rather than a business. The ICA is focusing more on business areas but still ACs in Europe are not interested in the ICA, in the cooperative movement. Specially in the EU there is the CAP, so in the European regions, ACs are more concerned about the CAP and they want to be more involved in lobbying and decision making in the EU.
>
> (ICAO rep)

A policymaker actually realised during an interview that "some of the cooperatives out there are in effect a slightly evolved version of a trade association" (UK policymaker). When this policymaker was asked about the social movement dimension of ACs, the following response was offered:

I would be thinking what are they doing in sharing best practice, guidance, education, commercial imperative [. . .] and maybe that does reflect a different view I have of what a farming cooperative is, compared to what I would normally as a centre-left politician say what a cooperative is. It is curious actually, there is a slight difference.

(UK policymaker)

This instrumentalist reading of cooperatives is something that the SAOS has also noticed in farmers in Scotland:

The primary purpose if it is a grain co-op, is to make the best return to the farmers for his grain. That is the reason why he wants to work together with other farmers because he thinks he can do that better and more effectively in a cooperative than doing it on his own.

(SAOS rep.)

Other participants also agreed with this view:

I think it is different. I think it is undoubtedly different. Most of the other cooperative forms are done by groups of people coming together on an individual basis and the individual members are not themselves commercial, they are coming together for an economic purpose. Whereas obviously the individual farmers are already kind of commercial entities before they come together to become members of a cooperative and that is a huge difference I think. So it is business owning a business, owning a business rather than individuals coming together to form a business [. . .] they have come together for quite specific economic reasons, rather than necessarily for having . . . there's never been much emphasis, certainly in this country, on the social side of cooperation, they have always been more interested in using cooperation as a vehicle to ensure their economic viability rather than necessarily because they think it is a great thing to do.

(Co-operatives UK rep.)

They see them very much as cooperatives that give them some commercial advantage, some edge, some ability to withstand the shocks long term.

(UK policy maker)

[M]aybe one of the things I'd say, historically in this country is that farms, and even independent farms have been bigger and wealthier and more commercially viable that they've been in the continent. So when you're talking about agricultural coops in the continent you're probably talking about smaller, poorer farmers whereas especially in some of the sectors in this country, the farmers were quite big anyway, when you're talking about farmers you're not talking about peasants, you're talking about fairly wealthy people in society anyway. [. . .] most of the cooperatives that grew up in breadth in the nineteenth century were really there for poor working class people trying to meet very basic needs. Agricultural cooperatives grew for wealthier people to meet rather different economic needs.

R: So both are meeting needs, it's just a very different set of needs?
I: Exactly.

(Co-operatives UK rep.)

[O]ne of the key things that a cooperative does, it is a form of economic democracy . . . in that sense the social benefit from the cooperative is not necessarily primarily in the work that they do, it is on how the risks and the rewards are shared around. So that the social benefits of cooperation are around economic democracy and participation and more equitable sharing of wealth and distribution of wealth. Whereas the Community Interest Company, the primary social purpose should come from the work it does, whereas for a cooperative is around how the surplus of the work it does is shared around.

(Co-operatives UK rep.)

This pursuit of growth regardless of the impacts on members and nature reflects both anthropocentric and Western epistemological biases, as noted by one interviewee:

If you want to portray [the biases] in today's cooperativism, that is, disregarding of other cultures, only wanting to grow and practise agricultures that abuse nature. Then, cooperativism also, it is not a separate entity. What we have is the fruit, the reflection of the model of society, of course.

(*Soberania Alimentaria* magazine rep.)

An ethnocentric bias also emerged from the data. From the 1950s onwards, the ICA became an active promoter of cooperatives as a development tool in low- and middle-income countries, but critics have pointed out that the imposition of the cooperative model abroad had a paternalistic, Eurocentric approach (Rhodes, 2012). Links between colonialism and globalisation are a serious issue when top–down ACs are promoted in less developed countries by both the EU farming industry and governments (Berthelot, 2012). This quote from the ICAO representative reflects just how current the colonialist baggage is:

Actually networking wise, they [large ACs] are focusing on their own business, on the CAP. But individually they are also interested in supporting and providing some benefits to the agricultural cooperative movement in developing countries but their connection is with their past colonised countries. The UK has some programmes with countries in Africa that used to be their colonies. In Norway and other regions they also have connections.

(ICAO rep.)

These new ACs are created through top–down approaches, negating the intrinsic democratic and bottom–up spirit of the cooperative movement:

> [I]n developing countries it is mostly top–down. It is quite normal these days; [. . .] top–down is more efficient because they get some training. [. . .] It is very difficult to have more bottom–up processes in developing countries. They do not understand the system of cooperatives, because a cooperative is not just a business, it is a democracy. But in many developing countries, farmers and proprietors don't think they are the owners, they are just users, but in the cooperative they are the owners and users of the business.
>
> (ICAO rep.)

This use of cooperatives as an economic and development policy tool imposes Western models of cooperation on other countries, reducing the space and ability of those being "developed" and forced to cooperate to define their own understanding of cooperation in their own terms (Böhm et al., 2010; Illich, 1976). This approach is also creating a path-dependent development model of ACs and farming methods in those countries, as well as reliance on institutions and different levels of inputs from more "developed" countries (Böhm et al., 2010).

Other authors have provided evidence of this bias in developing countries, where poorer cooperative members, and especially women, suffer inequalities in ACs dominated by larger landowners and foreign interests (GeoSAS, 2012; Griffiths, 2012; Huggins, 2014). Huggins's study of land grabs in Rwanda has also uncovered the top–down government approaches and the unaccountable control that elites wield over many ACs in the country, which are clearly dominated by larger landowners and foreign interests (Huggins, 2014) and in which poorer cooperative members are finding it increasingly difficult to gain access to fertile plots. Additionally, in a reflection of the androcentric bias, women were found to be less likely to be able to afford the minimum monetary contribution for cooperative membership and, therefore, unable to join ACs (GeoSAS, 2012). Rwanda is not an isolated case, and these issues appear also in fair trade cooperatives (Griffiths, 2012). Huggins found that farmers no longer had control over the crops cultivated or inputs invested:

> The policy of regional crop specialisation, in tandem with a policy of "encouraging" smallholders to join producer cooperatives that is often implemented coercively, restricts the ability of farmers to make their own decisions regarding crop choice, investment in inputs and marketing.
>
> (Huggins, 2014:372)

These concerns were shared by farmers interviewed as part of this research: some mentioned how their ACs would strongly suggest to members what varieties of cherry they should grow, always pushing for varieties that produced a bigger cherry, even if the flavour was worse than that of traditional local varieties and even if the new varieties were not adapted to the local climate and pests. Some interviewees also mentioned they found themselves in a lock-in situation with regard to their growing methods. Farmers wanting to become more sustainable mentioned barriers to transforming to organic production, as their ACs could not or would not accommodate small volumes of organic produce, as it did not make business sense.

Furthermore, European ACs are increasingly looking for new markets to continue their expansion, as the European food and drink market is perceived as a mature and highly saturated market, where significant growth is no longer possible (Bijman et al., 2012). This focus on globalisation and on using ACs as a way of concentrating production in order to get food out of home markets and into export markets severely clashes with the efforts of the food relocalisation movement.

This section has provided findings that challenge the widespread assumption that ACs are part of a global movement aiming to transform society into a more cooperative one. The movement is not homogeneous, at least not where ACs are concerned. It has also revealed the ethnocentric bias of the more corporate practices of ACs in Europe that are exporting a neo-liberal version of the cooperative model to developing countries.

In summary, ACs are increasingly instructing members what crops to grow, what varieties and what growing methods they should use in the framework of the dominant productionist paradigm (Lang and Heasman, 2015). Civil society groups have warned of the level of power concentration and the intensive monoculture approaches that ACs are instructing their members to adopt in order to standardise and increase production, moving away from the original cooperative objective of transforming the world, to merely adapting to and reproducing the system in which they operate (*Soberania Alimentaria*, 2013).

The future is polarised

As part of all the interviews, participants were asked about their views on the future of ACs and how they believed ACs' relationship with the wider food system is likely to develop. Most participants warned of a growing polarisation in the sector: at the one end, some large, powerful ACs will continue to get bigger and follow internationalisation strategies. At the other end, some small ACs will find niches in which to thrive. The fate of those in the middle, according to the participants, is uncertain, but options seem few: they will have to merge, grow or get out. When discussing the dangers of pushing for the creation of really large players, one of the policymakers recognised the following:

> It is true that to form, well, they can't be called cartels, but form people that powerful at all levels, we need to be very careful about this, of course. Luckily, that is not the task of this ministry; it's the task of the General Secretariat for Commerce, to ensure this is uniform.
>
> (Policymaker)

This quote shows the extent of the lack of policy integration in cooperative policy. Additionally, this scenario of polarisation seemed inevitable to many participants:

> Most probably there will be two trends. There will be first, further con-solidation, and I sincerely hope, that for instance in Spain, in the case of olive oil for instance, we see some consolidation, because we see some major opportunities for the olive oil producers and their cooperatives in this respect [. . .] So I have a reason to believe that in the South we will see bigger cooperatives especially in certain sectors where they would have a stronger position in the market place, would influence their economic returns, have better access to export markets or EU markets, put money in research and development and so forth. We already have this for instance in fruit and veg, we have some major cooperatives in there, we have some major cooperatives in wine, we must probably also see some further con-solidation. That is one trend [. . .] but in the North, we might probably see also the trend, and we already see this, the trend of setting up new cooperatives that bring in this locality to a certain extent.
>
> (Cogeca rep.)

At this point, the interviewee started to explain the case of rural water coopera-tives in Finland bringing together farmers with other rural actors as an example:

> Water cooperatives which is a new form of cooperative sector [. . .] mean-ing that individuals, not necessarily farmers, farmers only, but including farmers, they join forces with the local community and set up a coopera-tive where they organise their water supply for instance, or drainage, or sewage, or whatever, so that we see [. . .] these new forms or new coop-eratives also coming to Northern member estates or all member estates in fact, where the cooperative form of enterprise could give some assistance to certain developing sectors in better servicing, especially in rural areas.
>
> (Cogeca rep.)

The Spanish policymaker also agreed that the future of the AC sector has a high probability of becoming very polarised:

> Extremes are going to fall off the CAP and off everything because they are no longer going to receive subsidies. There is a study from FEGA, which is our body in charge of administering EU funding payments; a subsidy administrative file costs 300 euros per year in time costs, civil servants and so on. As we are working towards "capping" which is to cut from the top and from below, [so] beneficiaries of more than 300,000 euros are going to be taken out and beneficiaries of less than 300 euros are going to be taken out.
>
> (Policymaker)

A civil society participant put this polarisation into historical perspective as a normal trend in the cooperative movement:

> I think if you tracked cooperatives, sort of trends over time, you'd see, all across the world, if this is a graph of members, certain cooperatives and over time, you will see these kind of like humps come and go, and all of this, all the time, like there are big movements that come and go and come and go . . . I think partially reflecting changes in society, but there are sometimes massive strategic mistakes made by big cooperatives, or by sectors or by policymakers that have led to movements being pretty decimated. I think if you were to look at movements internationally now, and we've had conversations with people about this, it is that people feel there are two co-op movements now happening; there's the existing and established and the new and the radical, but we probably should not forget that the existing and established were probably the new and the radical a few decades ago.
>
> (Plunkett Foundation rep.)

This dual character of the cooperative movement was also found by this research in the AC sector. Chapters 8 and 9 will cover the "new and radical" cooperatives that the above quote is referring to. A key question that emerges is, if many ACs disappear as a result of the growing polarisation, will there be a tipping point from which the movement won't recover?

> Perhaps the fear is with regards to what you were talking about, to watch out, and do not create monsters, or transform ourselves into the US, which is to have four big ones and no small ones. We would lose the European essence, right? But oh well, let's see.
>
> (Policymaker)

Besides, as discussed in the previous and current chapters, the future for middle-sized ACs is not looking promising:

> [T]he reality is going to make these in the middle disappear. I think in Spain we are going towards polarisation, because for sure we are not going to create an agrarian middle class.[10] I can't see it, but hey, I hope I'm wrong. But no political or economic element leads one to think this is leading towards homogenisation. These ones [the middle ones] will get out of the market, and these ones are going to keep going up [the big ones]. [. . .] If we are victims of our own success it is not a good thing. We might have to modulate this policy in a couple of years [. . .] if you look at what is happening in the sector, it is true that is a bit polarised, but it happens everywhere, egg, milk, vegetables, oil . . . there are fewer companies remaining, but those that remain are more competitive and sell more, it's logical.
>
> (Policymaker)

As discussed in reference to the new Spanish law that compared Spanish ACs to Dutch ACs as a justification and inspiration for encouraging mergers in the sector, policymakers are also looking for inspiration in other European countries to reduce the number of CAP beneficiaries and support rural development:

> Look, there were 904,343 CAP beneficiaries in Spain in 2012. If we cut a bit from the top and a bit from the bottom . . . France I think has 300,000 PAC beneficiaries and they are much more competitive than we are and they have rural development, and rural regions, and crafts . . . England has commercial farms much bigger and they don't dismiss crafts, rural development, one thing should live alongside the other.
>
> (Policymaker)

When commenting about the concentration of food retailing in the UK, with four supermarkets accounting for roughly three-quarters of all food sales, the same participant pointed out that, in Spain, it "is very, very similar, and every day it is getting worse, much worse". The policymaker also spoke about Brussels's limitation to promote the legal cooperative form as such:

> R: Do you think from the EU they're going to bet on the PO model or the cooperative model?
> I: PO.
> R: And the legal model of cooperative?
> I: Brussels doesn't care about that. The OCM the aim is to concentrate produce.
> R: Regardless of the legal model?
> I: The thing is they can't, they don't have powers, they can't get involved in the society model that one might have. Then through the OCM[11] regulation they introduce the compulsory model of POs copying the model of milk, the milk pay packet, and the rewards more specifically through the 1st and 2nd pillars [. . .] Brussels cannot legislate the company law of member countries, so they do it through produce, no through how you manage to concentrate that produce.
>
> (Policymaker)

Another concern about the future that might become an imminent problem for ACs – especially in the UK, as the Conservative government is strongly pushing for big data in farming – is the risk of co-optation of agronomic data, raising questions over whether the owner of the data is the AC or the farmer member (Janzen, 2015; Ferris, 2017). Anecoop, the Spanish case study that will be analysed in Chapter 7, is already using a type of software called Visual 3.0. This software claims to help farmers submit CAP forms with fewer errors and have a register with all the requirements of the main certifications, such as Global Gap v. 5 and Integrate Production (Fresh Plaza, 2016). In the UK, the Conservative government is strongly pushing for a big data approach that, it claims, will improve the farming industry, and it has invested £11.8 million in a new "Agrimetrics Centre" to promote use of big data and analytical tools across the food system (Innovate UK, 2015). Issues of misuse, power and ownership of big data remain to be tackled, plus civil society voices claim the focus on data distracts policy action from more urgent and pressing issues in food and farming.

Conclusions

Many ACs in Europe are following commercialisation, internationalisation and concentration strategies in order to remain competitive in global food markets. However, this is coming at a price. Current measures of success are based on short-term profits rather than long-term gains. Evidence presented suggests many ACs are forgetting/have forgotten their radical and transformative roots and their link to the wider cooperative movement. This evidence has been categorised in eight different themes: legal dimension, governance, issues related to sustainability, lack of cooperative education, the importance of language, the effects of path dependency, (lack of) identity as a social movement, and the risks of a polarised future in the AC sector. Those large ACs ranked as the most successful often lack meaningful member participation.

Data have revealed how, with industrial agriculture and globalisation, the alterity of ACs is being diluted, and ACs are losing their identity and role as expressions of alternative food initiatives at the farming level to empower food producers and tackle power imbalances in the food system. The review of the findings presented in this chapter suggests that a wide range of multilevel factors are encouraging the corporatisation of ACs in the EU:

- cooperative level: lack of continuous education of members on the history, meaning and value of their cooperative and the wider cooperative movement;
- national level: policies aiming to increase the profitability of farming as an economic sector; clear support for consolidation of the sector and larger farms;
- EU level: CAP pushing for PO model to concentrate demand and facilitate trade in the Common Market and beyond;
- global level: the introduction of global agricultural trade and increasing demand from emerging economies are also pushing ACs both to become bigger players to be able to compete in their national markets and to export to foreign markets and meet the demand needs of big suppliers.

Notes

1 For this interviewee and all following interviewees who are cited, please see the table in Appendix III.
2 ISO is an independent, non-governmental international organisation with a membership of 161 national standards bodies. ISO sets out the requirements for a food safety management system and can be certified to.

3 Mercadona is the largest food retailer in Spain, recently criticised for sourcing food from North Africa instead of supporting Spanish farmers.

4 Integrated production refers to a sustainable farming approach that produces food and other products by maximizing the use of natural resources and regulating mechanisms to replace artificial inputs.

5 BOE is the Spanish Official State Bulletin in which government communications are published.

6 The participant was referring to the financial crisis that has depressed the Spanish economy for a decade now since it started in 2007–8.

7 The participant refers here to the repression and dissolution of cooperatives during the Franco dictatorship. Cooperatives had imposed on their board of directors a representative from the regime and one from the Church. This topic was discussed in detail in Chapter 2.

8 NIF is the Spanish tax ID number.

9 Fagor was one of the most famous workers' cooperatives, part of the group Mondragon; it went bankrupt and was auctioned after Mondragon decided not to rescue it.

10 In Spain the "middle class" is the working class; upper-working class can be considered the equivalent of the UK middle-class label.

11 OCM stands for Organización Común de Mercados, which refers to the EU's common organisation of the markets in agricultural products.

6   Country cases – UK and Spain

From workers' unions to the European Union

Why are Spain and the UK two interesting case studies of agriculture cooperative sectors to compare?

So far, we have discussed the huge diversity of ACs that exists in Europe. Nevertheless, the European Commission has identified some trends: in northern Europe, ACs are fewer, larger and more specialised; in the south, the sector is, on average, still quite atomised and localised.

As a way to delve deeper into this diversity, it is worth considering two case studies, one from the north and one from the south of Europe. The UK and Spain were selected, and the choice was based on a number of rich historical and socio-economic differences and present similarities that are discussed below.

The idea of country was applied loosely, even where important intranational differences exist (Hantrais, 1999). This point is especially relevant in the case of Spain, where nationalisms and regional identities are very strong factors in some of the case studies based in Catalonia and the Basque Country; the UK will be considered one country for the purpose of this research, although formed of several countries itself. First, let's consider the differences:

The UK was the first country to:

- have a formalised cooperative (see more about the Rochdale legacy in Chapters 1 and 2);
- undergo the Industrial Revolution, with its associated transformation of society;
- transform its agriculture into a highly industrialised economic sector;
- consolidate farming (only second after Denmark; by the 1980s, the UK had the largest farms in Europe).

Additionally, in the UK, agricultural cooperation:

- developed less and more slowly than consumer and industrial cooperation;
- became a tool for economic protection and of top–down measures.

Furthermore:

- Agriculture accounts for 0.5 per cent GVA/GDP and 1.4 per cent of employment (DEFRA, 2012).
- There are 434 (down from 621 2 years earlier) ACs in the country, providing approximately 8,000 jobs (Co-operatives UK, 2017).
- Agricultural cooperation does not receive funding from the UK government (only partially in Scotland), with vertical cooperation in the supply chain being more prominent.

In contrast:

- Spain was one of the last 17 Eurozone members to industrialise agriculture (Simpson, 2005).
- Early agricultural cooperativism in the late nineteenth and the first three quarters of the twentieth century, became a control tool used by the Church and the Fascist regime to counterbalance communist and anarchist ideologies, which also promoted cooperatives.
- Currently, Spain has the second strongest cooperative economy in EU, accounting for 2.2 per cent GVA/GDP and 4.3 per cent of employment.
- Spain has 4,000 ACs providing approximately 100,000 jobs (OSCAE, 2015).

At the same time, these two countries also share some similarities in their approach to AC policies:

- Conservative governments promoting the cooperative economy: At the time this research started, Spain was governed by the Partido Popular, a Christian democratic and conservative party. In the UK, the Coalition government, led by the Conservatives and the Liberal Democrats, was pushing the concept of the Big Society. While the research for this book was being carried out, many political changes took place in both countries. In Spain, two new political parties emerged and are now key actors in the political arena. In the UK, the Conservative party won a majority in May 2015 after 5 years of their coalition with the Liberal Democrats. It is still unclear what the Brexit vote in June 2016 will mean for UK farming and ACs.
- Both countries have undergone recent cooperative legislation changes: Spain introduced a new law to promote mergers and acquisitions in the AC sector in 2013. In 2014, the Coalition government in the UK approved the Co-operative and Community Benefit Societies Act 2014, bringing cooperatives on to the agenda and combining 17 pieces of different legislation that had not been updated for decades.

History of agricultural cooperation in the UK: from workers' unions to the European Union

Since the very beginnings of the movement in eighteenth-century Europe, food has always been a core element of cooperativism (Burnett, 1985; Birchall, 1994;

Garrido Herrero, 2003; Rhodes, 2012). The connection with food was key especially in the development of the UK's cooperative movement, as new urban workers who had lost the means to grow or forage their own food started to explore different avenues to secure sufficient, safe and affordable food provisioning. In 1851, just a few years after the opening of the Rochale store and also around Manchester, the first worker cooperative farm in the UK, Jumbo Farm, was founded and it operated for 10 years (Birchall, 1994). Neither Jumbo Farm nor the Rochdale store were perceived by their originators as an end in themselves, but as means to self-provide food and funds for bigger industrial cooperation projects that could pave the way for a deeper and wider transition towards a new model of society aligned to the cooperative principles (Oakeshott, 1978; Fairbairn, 1994). The Rochdale Pioneers did not buy or rent land to employ members until 1896. The Co-operative Wholesale Society (which developed from the original Rochdale store and is now named the Co-operative Group) began farming in June 1896, when it bought the 742-acre Roden Estate in Shropshire to grow fruit for its jam factory. For decades, the Co-operative Group was the UK's largest agriculture employer and landowner, until, in summer 2014, the group sold its 7,000 hectares of land, 15 farms and about 130 residential and commercial properties to the Wellcome Trust for £249 million, in an attempt to cover its debts (Co-operative Group, 2014). The Wellcome Trust said it would be a responsible landowner but did not sign up to the Co-operative Group's ethical guidelines (Moulds and Treanor, 2014). A campaign by a network of NGO members of the UK Food Group asking the Co-operative Group to sell the land to small- and medium-scale ecological farmers or cooperatives of farmers went unheard in favour of a single, simpler financial deal. This sale ended the farming arm of the first formalised retail cooperative in history, still one of the largest consumer cooperatives in the world today.

Paradoxically, ideology, rather than financial deals, was what fuelled the original Rochdale Pioneers (Oakeshott, 1978; Fairbairn, 1994). But these early cooperative experiments did not come out of a blank political and philosophical sheet. By the 1850s, a clearly defined cooperative ideology emerged out of the convergence of many influences (see Table 2.1 in Chapter 2). The first cooperative ideas date back to feudal times, when peasants held land in common.

More modern influences include the Chartist movement, Adam Smith's concept of the "moral economy", Cromwell's Commonwealth[1] and the concept of "commonweal" (reinforcing the idea of common resources and the welfare of the public). Robert Owen added moral and community dimensions to the cooperative thinking that was developing at the time. Although still in an authoritarian manner, Owen put his theories into action in an attempt to improve the well-being of the working class through his social experiments in New Lanark (Scotland) and, less successfully, with the New Harmony community in Indiana, in the US (Birchall, 1994; Rhodes, 2012). William King and the Rochdale Pioneers added the democracy, self-help, self-management and mutuality elements to the movement (Rhodes, 2012). Class consciousness also fostered the development of the cooperative movement.

According to Thompson's account in *The Making of the English Working Class*, Thomas Paine's work *Rights of Man* (2008/1791) was the foundation of the working-class movement and the beginning of a way of thinking that acknowledged the rights of man over the church and flag (Thompson, 1991; Glasman, 1996).

The enclosure process also unintentionally manufactured a "radical agrarianism" phenomenon that spread from the 1770s until the mid nineteenth century and was based on the emergence of "land consciousness", parallel to the development of "class consciousness" (Glasman, 1996). Despite happening in a time of urbanisation and industrialisation, both the radical working-class politics of Chartism's early socialism and the cooperative movement shared conceptions of nature, land and the environment as their common denominator, fighting against the commodification of society and nature brought about by the enclosure process (Glasman, 1996).

Robert Blatchford, famous for his social nationalist ideas and his book *Britain for the British*, refuted the claim that it was cheaper for the UK to buy food on the international market than to grow it nationally in his chapter "Can Britain feed herself?" (Blatchford, 1902). The title of this chapter has been reused several times to name books and articles asking the same question of the late twentieth- and early twenty-first-century UK (Mellanby, 1975; Fairlie, 2007; *The Land*, 2012). But, at the time Blatchford wrote his book, investment was being redirected from farming into the development of industrial activities, leaving behind acres of uncultivated land and millions of idle unemployed workers. Blatchford discussed the issues of "intensive cultivation", the difficulty of accessing land and food security in times of war as well as of food availability for and affordability by industrial workers on very low wages (Blatchford, 1902:116). Blatchford presented cooperatives as the living example of working socialism (Blatchford, 1902). Those early concerns over the commodification of food, the industrial appropriationism of "natural" food production methods and resources and the choice between cheaper imports and domestic production are still very present today and represent core debates in the AC sector.

However, despite having a strong ideological basis, it was only after the Rochdale store was established that political lobbying for a specific cooperative legislative framework took off in Britain (Rhodes, 2012). It was not until 1852 that the legal cooperative form entered the law for the first time in history (Zeuli and Cropp, 2004). The Christian socialist movement and the liberal economist J. S. Mill partnered to lobby for favourable legislation to remove barriers to saving for the working class, resulting in the 1852 Industrial and Provident Society Act (also known as Slaney's Act); it is interesting to note that the legislation did not include the word cooperative in its name; additional measures followed in 1893, 1952, 1961 and 1965 (Coates and Benn, 1976).

Cooperativism in agriculture continued to develop after the initial Jumbo Farm experiment. Led by Edward Owen Greening, the first supply AC in England was formed in 1867. Called the Agricultural and Horticultural Association (AHA), its main function was to bulk buy seeds and fertilisers for members, in an effort to combat both high prices and adulteration of inputs

(Rhodes, 2012), a function that it carried out successfully for 50 years until it closed in the middle of World War I (WWI; Morales Gutierrez et al., 2005). Greening was a promoter of cooperation who knew many Owenites (followers of Robert Owen), including Holyoake, and he went on to play a key role in setting up the ICA. Greening counted on the backing of many Catholic socialists. This type of socialism developed after the failure of Chartism and when the conflict between Christian principles and business activities became obvious. The influence of religious groups in the development of the cooperative movement is worth noting. Quakers had an active role in promoting cooperation and lobbying for a suitable legislative framework. However, it has been argued that it was not so much that the religious beliefs of Quakerism led directly to entrepreneurialism and the setting up of cooperative businesses, but rather that the network in which the Quaker community lived fostered trust and exchange of information, as well as of key financial resources (Spear, 2010). Moreover, many Quaker families lived in Manchester, a region in which many cooperative enterprises developed, and the "politically neutral" world of business was an area in which they could freely participate and thrive (Spear, 2010).

As mentioned earlier in this chapter, fighting adulteration was a shared objective of both early consumer and agricultural cooperatives. Given Greening's circle of influence and networks, it is not surprising that the AHA replicated the model of existing successful consumer cooperatives (Knapp, 1965). Greening introduced a more formal model than previous attempts to empower farmers through cooperation had done, and the AHA operated for almost 50 years (Knapp, 1965).

Legal factors such as inheritance laws and the transformation of public land by rich landowners in the late eighteenth century through "enclosures" triggered the acute concentration that defines UK farming today and acted as a deterrent to widespread cooperation in farming. The reduction in the number of both farms and farmers happened at the same time as agriculture was being transformed into an economic and commercial sector, with farmland being converted into increasingly large, more efficient units. This transformation noticeably decreased farmers' marketing needs and motivation to cooperate. In contrast, as will be discussed in the next section, Spain's numerous smallholdings had very different technical and marketing needs, and farmers felt a more immediate need to pull their produce together to be able to access markets.

Back in the UK, the more popular retail cooperatives evolved to serve a classless consumer and left class conflict out, as everybody in urban areas was a consumer of certain goods and shared the need to shop. Two factors – the focus on consumption fostered by the Industrial Revolution and the distancing of the population from previous more food-self-sufficient lifestyles, replaced by a reliance on the market – strengthened the importance of consumer over producer cooperatives. Development policies for the colonies were, however, quite the opposite: in colonised countries, the focus fell on fostering cooperation at the production level to secure the food imports required by their increasingly industrialised mother country (Rhodes, 2012).

In Britain, government regulation of exports and imports of corn was well established long before the nineteenth century (Schonhardt-Bailey, 1996). But, in the years leading up to 1846, when the Corn Laws were repealed, the conflict between the declining landed aristocracy and the rising manufacturing and export industrialists escalated. The latter wanted to repeal the Corn Laws to facilitate the export of British products by improving overseas grain buyers' foreign exchange and, thus, their purchasing power. Industrialists also expected the repeal would introduce more competition in the sector, resulting in access to cheaper food for their workers, who were considered another "input" needed to continue the development of the Industrial Revolution. Agricultural elites, by contrast, wanted to maintain the inflated prices that the Corn Laws had offered them so far. Robert Peel, prime minister at the time, decided to repeal the Corn Laws to appease the industrial and working classes (Schonhardt-Bailey, 1996). Peel reduced duties for many products, both manufactured and food, and offered some incentives to the landed aristocracy to lessen their financial loss, including the consolidation of a major road network and loans at low interest rates for agricultural improvements (Schonhardt-Bailey, 1996). This development into free trade fuelled the adoption of high farming (modern growing techniques) during the period between the 1840s and 1880s (Prothero et al., 1961).

By the end of the nineteenth century, the importance of farming to the economy had already dramatically decreased, and the sector employed no more than a quarter of the population (Hobsbawm, 1999). As a result, a significant proportion of the redundant peasant workforce moved to cities, reducing the amount of labour available to rich landowners. This higher demand for land-workers raised agricultural wages for almost the first time in history, as well as eliciting an interest in labour-saving farming methods (Hobsbawm, 1999).

Beatrice and Sidney Webb defined agriculture at this time as "the most remarkable decline of an unregulated and self-supporting industry" (Webb and Webb, 1897:761). It is interesting to notice how, as early as 1897, the Webbs already referred to UK agriculture as "another industry". Wealthy landlords acted as "shock-absorbers" of crises for farmer tenants and kept their political temperature low (Hobsbawm, 1999). In contrast, Spanish peasants did not have an alternative industrial sector that could absorb them until much later on (Garrido Herrero, 2003). As a result, the number of ACs in the UK remained insignificant at the beginning of the twentieth century. Authors have listed further factors that caused this poor development, including the traditional attitude of conservative farmers (Rhodes, 2012) and resistance from landlords (Rayner and Ennew, 1987), as well as the fact that markets were inundated with cheap imports and land value decreased in parallel to the decreasing importance of farming to the national economy (Rhodes, 2012). However, this picture changed at the beginning of the twentieth century, when many ACs started to develop, and each country in the United Kingdom established its own body to encourage the formation of new cooperatives (Knapp, 1965).

Towards the end of the nineteenth century and in the first half of the twentieth century, Horace Plunkett, the son of an Irish peer, became a key figure in the farming cooperative movement in Ireland (where he led efforts to support small-scale Catholic farmers) and later also in England. In 1894, Plunkett supported the creation of the first federation of cooperatives in Ireland under the name of the Irish Agricultural Organisation Society (IAOS). He also stimulated England's interest in cooperation, resulting in the establishment of the British Agricultural Organisation Society in 1900 for England and Wales. Plunkett went on to found the Plunkett Foundation in 1919, a charitable organisation still supporting rural communities today. Similar to the situation in Spain, it is interesting to highlight the more agricultural character of early cooperation in Ireland, where Plunkett's work started, compared with the consumer focus the movement had in the UK. Similar stages of development facing the two countries and a shared Catholic ethic meant that agricultural cooperation in Ireland resembled more that in Spain, as will be discussed in the next section.

In 1901, the British Society merged with the National Agricultural Union and was registered, under the Industrial and Provident Societies Act, under the name of the Agricultural Organisation Society (AOS). The cooperative was designed to:

> secure cooperation of all connected with the land, whether as owners, occupiers or labourers, and to promote the formation of Agricultural Cooperative Societies for the purchase of requisites, for the sale of produce, for agricultural credit, banking and insurance, and for all other forms of cooperation for the benefit of agriculture.
>
> (Knapp, 1965:12)

It is worth pointing out that the AOS was initially supported by donations and member subscriptions, a reflection of the self-help character of early agricultural cooperativism and its independence from government.

The National Farmers Union (NFU) also supported ACs, especially after the dissolution of the AOS. However, as Joseph Knapp has documented in his detailed account – one of the only ones – of the history of agricultural cooperation in England, the NFU failed to give full recognition to agricultural cooperation as an independent self-help business activity of farmers (Knapp, 1965). In fact, representative cooperative organisations and the NFU had a conflictive relationship, competing to be the voice of farmers from the late 1920s until the late 1960s, when the UK started to prepare to join the EU (Morales Gutierrez et al., 2005). During the first half of the twentieth century, the cooperative movement in the UK did not experience much interference from the government, or support. During the interwar period, however, a different policy context was created by the government and the National Farmers Union with the introduction of compulsory marketing schemes and efforts to further modernise farming (Morales Gutierrez et al., 2005). During this time, the UK government introduced the Agricultural Marketing Acts of 1931 and 1933, probably as a response to the slow development of ACs. These Acts announced the creation of four marketing boards for the milk, bacon, hop and potato sectors (Morley, 1975; Rayner and Ennew, 1987). Other boards were established after WWII. Most of the boards, with the exception of the Bacon Board, survived until the mid 1990s. The existence of the marketing boards within many sectors has been identified as the main factor that removed the perceived need for bottom–up cooperative action in UK farming (Rayner and Ennew, 1987).

The post-war period started the era of the Federation of Agricultural Cooperatives in Great Britain and Ireland (FAC). The FAC was formed in 1949 as a central body to coordinate the work of existing federal agricultural organisations throughout the British Isles. The FAC consisted of the Agricultural Central Co-operative Association (ACCA), the Welsh Agricultural Organisation Society (WAOS), the SAOS, the Ulster Agricultural Organisation Society (UAOS), the IAOS, the Agricultural Co-operative Managers Association (ACMA) and the Plunkett Foundation for Co-operative Studies (Knapp, 1965). During the 1950s, the FAC acted as a political lobby but also provided training, managed relations between cooperatives and marketing boards, collected statistics, coordinated trade between cooperatives in the various member countries and promoted international cooperative relations (Knapp, 1965).

By the 1960s, the AC sector in England had already started a process of consolidation that continues today, creating fewer but larger cooperatives competing for a higher share of their markets. This could be seen as a response from the farming community to the fast development of and level of concentration in the processing and retail sectors. In 1965, Joseph Knapp was commissioned by the then government-funded ACCA to carry out a review of agricultural cooperation in England. Knapp highlighted the difficulty of achieving economies of scale while still engaging farmers as members and owners rather than mere customers of cooperative services (Knapp, 1965). The report was commissioned because UK farmers were concerned with the fast degree of vertical integration of English food production and the exponential growth of industrial farming units, specially "giant beef-lots and egg production units" (NFU, 1964, in Knapp, 1965). In this report, lack of research on agricultural cooperation (at government, university and industry levels) was identified by Knapp as one of the major weaknesses of the sector (Knapp, 1965).

The 1967 Act was part of the preparations to get the UK ready for its entry to the European Community by bringing changes to existing support mechanisms. The most dramatic measure, which went on to shape the growth and development of the AC sector, was the move from compulsory marketing boards to voluntary cooperation needed to comply with EC competition law (Davey et al., 1976). The Agriculture Act from 1967 called for the creation of the Central Council for Agricultural and Horticultural Cooperation (CCAHC), which boosted cooperation in agriculture. However, a few years later, the Agriculture Marketing Act of 1983 called for the creation of Food from Britain (FFB), which subsumed CCAHC and operated until 1993 (Environment, Food and Rural Affairs Committee, 2010; WAOS, 2011). With the creation of FFB, which will be discussed later in this chapter when examining the AC sector in Britain today, government support for ACs in the UK (with the exception of Scotland) started to focus more on "collaboration" than "cooperation" per se (EFFP, 2004).

History of agricultural cooperation in Spain: anarchism, Francoism and the Catholic Church

Agricultural cooperativism in Spain developed from informal agrarian associationism, with agrarian unions being the first cooperative experiments in the country (Campos and Carreras, 2012). At the social and political level, two factors awakened peasants' awareness of their rights and of class and power inequalities: the introduction of universal male suffrage in 1890, and the diffusion of revolutionary ideologies during the Napoleonic occupation of the early nineteenth century and of Bakunin's anarchist ideas in the 1860s (Hermi Zaar, 2010). Moyano Estrada has explained how early associationism was framed in these wider ideologies (such as Marxism, socialism and Catholicism) that exceeded the limits of the agricultural world (Moyano Estrada, 1988 and 2001). Both the strongest anarchist union – CNT (National Workers' Federation) – and the socialist union – FNTT-UGT (National Federation of Land Workers) – fostered the creation of cooperatives for the communal cultivation of land as it constituted a channel for the collectivisation project that was core to their vision and campaigning (Paniagua, 1982).

The development of intensive monocultures introduced a narrower focus in the mission and purpose of ACs (Moyano Estrada, 1988). In Andalusia, in 1930, 60 per cent of male labour was still at the mercy of big landowners; high levels of unemployment and massive landownership disparities generated the strongest rural anarchist movement in Europe (Simpson, 2003). The empowering of peasants threatened the status and security of landowners (Planas Maresma, 2008). However, the high level of illiteracy across the country at the time hindered, not only the take-up of agri-technology, but also the development of a strong cooperative sector (Planas Maresma, 2008). With regard to economic factors, expensive and unsuccessful wars in South America as Spain attempted to retain control over its colonies in the new continent put it into a difficult economic situation. On top of that, other factors, such as several pests (e.g. phylloxera and mildew) that affected vineyards (that were already a cash crop at that time) and the end of the trade treaty between France and Spain (1892), only worsened the situation. In the middle of an economic crisis and lacking an alternative economic sector or urban areas keen to absorb their labour, Spanish peasants and small farmers found themselves without any bargaining power, in a state that could lead to either revolt or cooperation. However, neither of the two happened successfully.

The lack of social and financial capital meant that, although many attempts to start cooperatives took place, many were short lived (Garrido Herrero, 2003). Several authors (Garrido Herrero, 2007; Planas and Valls-Junyent, 2011) have linked their short lifespan to the poverty of the members: even when members pooled their resources together, the total amount of funds was still negligible and not enough for the kind of investment needed to set up a viable AC. This paradox highlights the tension between the "open door" character of cooperatives and the actual social and financial capital required to make them succeed. This tension was clear in early Spanish cooperatives, and their failure reflected the widespread deprivation and levels of illiteracy suffered by the peasantry.

From the top–down perspective, the first national cooperative organisation, namely the National Committee for Spanish Cooperatives, was set up in the mid 1890s (Bartlett and Pridham, 1991). With regard to agricultural cooperation, the Royal Decree of 1886 for Cámaras Agrarias (Agrarian Chambers) was the first to legislate for agricultural cooperativism, combining the Law of Associations of 1887 with the competence to develop cooperative activities, as well as advising government on the needs of the agricultural sector. However, only 30 associations were created within the first decade of the legislation being introduced. Some authors have suggested the low number was due to the fact these Agrarian Chambers encouraged a conservative and corporative model of cooperation that emphasised the traditional values of a romanticised agrarian social class in an attempt to reduce radical ideologies and interclass conflict (Garrido Herrero, 2003) and did not even recognise the one-member-one-vote cooperative principle (Planas Maresma, 2008).

Garrido Herrero (2003) has argued that the ambiguous 1906 law was ingeniously created to allow agricultural syndicates to be both syndicalist and cooperative. This allowed the top–down syndicates to claim they were working to meet peasants' needs, whereas in fact their main focus was on economic activities to cover the immediate and basic needs of the farming population to act as a deterrent to Marxist ideas/movements. Many peasants and smallholders developed a preference for the local agrarian syndicates created under the 1906 law over the conservative Agrarian Chambers, as the local syndicates offered them fiscal benefits. However, some of the more socialist, republican or anarchist agrarian cooperatives refused the government's model of cooperation and opted to remain unincorporated (Garrido Herrero, 2003). This dual cooperativism (co-existence of "rich" and "poor" cooperatives) has been documented as another factor that weakened the movement at the time (Planas and Valls-Junyent, 2011).

The most economically successful ACs during this early period were those supported by the Catholic Church (Garrido Herrero, 2003). Many authors have defined Catholic cooperatives as paternalistic, religious and anti–revolutionary and as having a double objective: "saving" peasants from the effects of anarchist and communist ideologies and providing economies of scale to help them avoid revolt-prone hunger (Simpson, 2003; Garrido Herrero, 2003; Gil, 2004; Giagnocavo, 2012). Set up to maintain the status quo (Hermi Zaar, 2010), no attempt to create an alternative cooperative economy, let alone society, was part of their agenda. As Simpson has described:

> [T]he absence of a democratic political tradition, which involved the active participation of smaller farmers, left the cooperative movement being controlled by the larger landowners, whose interests naturally differed.
>
> (Simpson, 2003:230)

Furthermore, Planas Maresma (2008) has identified four reasons big landowners were keen to join ACs and encourage the development of legally incorporated agrarian associationism in Spain:

• Cooperatives could serve as a tool to reinforce collective action as they had the potential to become the first forum for agrarian interests and the "agrarian class" consciousness that was emerging parallel to that of urban/industrial classes.
• ACs could serve as a tool to increase the social legitimacy of the landowning classes, which was starting to be questioned. Their active role in the apparently "classless" cooperative movement helped them project a fresher image linked to the agrarian modernisation process that was taking place.
• Landowners had growing concerns about their inability to stop the influence anarchist and communist ideologies were having on peasant populations and tried to reduce their effect. On the one hand, they attempted this by encouraging small farmers to join cooperatives; with this method, rich landowners stopped peasants from being politically inactive and took advantage of their numbers to increase their lobby power. On the other hand, by having them in their own cooperatives, peasants could be "controlled" and prevented from creating more radical cooperatives.
• ACs could also be a tool to fuel the technological development of their own land in an attempt to ensure more competitive yields and, thus, their rents.

In contrast, Planas Maresma (2008) has pinpointed two main factors that underlined peasants' motivation to join cooperatives dominated by big landowners: as a way to survive the crisis that the agricultural sector was experiencing, and to achieve a reduction of the costs associated with new growing methods and equipment. Peasants, according to Planas Maresma, joined ACs mainly for financial savings, and, if these were provided, their motivation or justification to complain about the managing style of the cooperative was reduced. Nevertheless, it is still worth noting that, though lower in numbers, cooperatives of other ideologies existed – for example, socialist and communist (Garrido Herrero, 2003) – especially after 1920, when the number of Catholic syndicates peaked and started to decrease (Hermi Zaar, 2010).

Official recognition of the cooperative model as such had to wait until the fall of the monarchy in 1931, when the first Act permitting the formation of cooperatives in any business activity or sector was introduced, which lasted throughout the Second Republic until the end of the Civil War in 1939 (Julia and Melia, 2003). With the outbreak of the Civil War in July 1936, their number increased rapidly, especially in Aragon and Catalonia, where the tradition of the anarchist movement was strong. In the countryside, many rural estates were expropriated, and cooperatives were formed to cultivate the land.

By 1937, more than 2,700 cooperatives had been created, but, by the end of the Civil War in 1939, Franco's regime brought this process of development to an abrupt end. The patrimony of the unions and cooperatives associated with the socialist and anarchist movements was seized, and, at the same time, many Catholic-agrarian organisations were incorporated into the new structure of the regime (Majuelo and Pascual, 1991). Some cooperators moved to South America, where they spread the movement's message and set up successful cooperatives (Birchall, 1994). Franco's authorities assassinated outspoken cooperators who were not able or decided not to leave the country (Giagnocavo, 2012). The addition of a representative of the regime and a representative of the Church to each cooperative board was made compulsory.

The final alignment of cooperatives to Franco's dictatorship was completed with the 1942 Act (Julia and Melia, 2003). This Act remained in force for the next 32 years and made cooperatives subject to rigorous legal control. Despite the development of the cooperative sector being inhibited by this legislation, it did not prevent the creation of Spain's most famous industrial cooperative, Mondragon, in 1956, which is still operating today. The creation in 1959 of a cooperative in Zúñiga (Navarre) that was run like a capitalist firm can be said to mark the birth of "commercial co-operation" in Spain (Majuelo Gil, 2001). Coupled with the development of industrial agriculture, cooperation started to move away from its original focus on people, to a focus on "enterprises" and business. The desire to develop an export sector for Spanish agriculture followed soon. In 1975–6, three experimental farms were set up in Almeria, fully financed by a cooperative bank. This was the beginning of the development of greenhouses and ACs in that region that has reached an inconceivable scale today, exporting around 80 per cent of the produce cultivated in the area (Giagnocavo, 2012). The case of Almeria and its close relation with the AC sector is discussed in detail in Chapter 5.

Franco's death and the return to democracy in 1975 brought about swift improvements in the cooperative movement in Spain (Barke and Eden, 2001). In 1977, the Freedom of Association Act was approved, and political parties, trade unions and employer organisations regained legal status. One year later, in 1978, the Spanish Constitution was enacted. A specific article of the new democratic constitution, number 129, explicitly stated the support the government should grant to cooperative societies by introducing appropriate legislation and also supporting workers to access means of production (Barke and Eden, 2001). Democracy brought a period of transformation and opening for ACs, and, with political devolution, legal responsibility for ACs was transferred to the newly established regional autonomous governments.

However, as discussed above, many leaders in the cooperative movement during Franco's period were closely linked to the regime at the local level, often also being part of the political elite (Majuelo and Pascual, 1991). Some authors have argued that this was the reason why the development of the more recent cooperative movement was disconnected from workers' union movements (Majuelo and Pascual, 1991). Additionally, during the democratic reform of the late 1970s and early 1980s, the new government encountered economic, bureaucratic and political barriers (owing to the ruling classes from Franco's regime wanting to retain power), slowing down the dismantling of the former regime's top–down cooperative structure. In the 1980s, political devolution changed the legal arena and policy context for cooperatives:

> With political devolution in the 1980s, legal responsibility for co-operatives was assumed by the newly established regional autonomous governments and, by 1985, laws had been passed by governments in the Basque Country, Catalonia, Andalusia and Valencia. By this time the co-operative movement in Spain was one of the largest in Europe, second only to Italy.
>
> (Barke and Eden, 2001:202)

Indeed, the number of ACs of both first and second degree (the latter are federations of ACs) grew exponentially, and, by 1984, the sector agreed on the need to create an umbrella body that represented them at the national level (Maté and Carlón, 1984). Soon after, the Confederación Española de Cooperativas Agrarias was created. In 1986, Spain joined the EU. Three months later, unhappy with the governance structure of the Confederación and its lack of solidarity with its member organisations, a group of regional second-degree ACs formed the Asociación Española de Cooperativas Agrarias (AECA), and others formed the Unión de Cooperativas Agrarias de España (UCAE; *El Pais*, 1986). The effort to make cooperatives more businesslike was increased when Spain joined the European Common Market in 1987; that year, a new national law that recognised cooperatives as businesses, as opposed to not-for-profit organisations, was introduced. In 1989, the two main representative bodies for ACs, AECA and UCAE, merged to establish a new Spanish Confederation of Agrarian Cooperatives (Cooperativas Agro-alimentarias, 2016). In March 1990, the confederation acquired legal status and was recognised by Cogeca in Brussels (Cooperativas Agro-alimentarias, 2016). In 2009, the confederation changed its name to Cooperativas Agro-alimentarias (Agri-food Cooperatives) to reflect the "modern" face of competitive ACs adapted to new markets (Cooperativas Agro-alimentarias, 2016). In the 6-year period immediately after the incorporation of Spain into the EU, an ongoing process of mergers in the agri-food industry took place, driving further consolidation in the sector (Serrano et al., 2015).

The most recent national cooperative law, from 1999 (still in force), emphasises the business side of cooperatives, contrasting with the ICA definition that also specifies cultural and social needs. In the current legal framework, agricultural cooperatives are seen as associations of farmers that pursue economic benefits, disconnecting the cooperative model from its traditional solidarity roots (Giagnocavo and Vargas-Vasserot, 2012). Similar policy developments are taking place currently too, as will be discussed later in this chapter.

Conclusions from historical background

This first part of the chapter has discussed the beginnings of agricultural cooperation in Europe during the nineteenth and twentieth centuries, focusing on two countries that, at the time, had very different socio-economic contexts: UK and Spain. Existing evidence shows that, through history, social change has been brought about by people organising themselves to correct a perceived inequity. Consumer and producer cooperatives are an example of this agency and soon became an international phenomenon. They quickly proved to be a successful mechanism to meet the needs of different groups, such as farmers and urban consumers.

In the UK, early ACs developed around price and quality concerns over food inputs during a time of fast technical innovations in agriculture and the development of a new industrial sector. In Spain, however, cooperation was used as a social tool by opposite ideologies: on the one hand, the Catholic Church promoted cooperatives as a more acceptable option than anarchism; on the other, the anarchist and socialist movements saw cooperatives as the democratic way of organising work that would free the working class. While Spain was struggling to maintain its empire, fighting revolutions in South America, Britain was relying on its colonies to feed its urban workers and was focusing on a different type of revolution: the industrial one. These two different starting points set the pace for the distinct paths of development that the AC movement followed in each nation.

The fact that the sector employing the bulk of the workers in the UK was industry, not agriculture (Birchall, 1994) meant that farmers' demands soon became irrelevant to the majority of the workforce. As a result, UK cooperativism flourished in retail, as the bulk of the working population was formed of industrial workers who had to buy their food and other goods. Peasants became consumers. In Spain, however, the movement had more political weight owing to the still high proportion of the population living off agriculture at the time and the spread of anarchist and communist ideas. The lower capital requirements of consumer cooperatives facilitated their success in the UK, both in terms of turnover and membership numbers. ACs, on the other hand, required higher initial investments, a constraint that reduced poor farmers' capacity to cooperate and shortened the longevity of many ACs.

Christian religious groups were promoters of cooperation in both countries, mostly ACs in Spain and consumer/industrial cooperatives in the UK (Quakers), but also key in the agricultural sector. In the UK, their influence was clear at the legislative and entrepreneurial level. In Spain, the Catholic Church used cooperatives as a poverty alleviation tool, but also as a control tool to stop peasants from rioting against inequalities. UK Quakers had a more empowering idea of cooperatives compared with the poverty relief role that characterised ACs organised by the Spanish Catholic Church. Catholic ACs were designed to improve peasants' conditions in an effort to prevent revolts and the possibility of a French-style revolution (Simpson, 2003). Initially, in the UK, ACs had a focus on self-help and democracy, which then started to weaken in the first decades of the twentieth century when the sector had a complicated relationship with the NFU. Later, the compulsory marketing boards also removed the vision and need for bottom–up and voluntary coop-erativism. This rigid version of top–down compulsory agricultural cooperation was taken to the limit in Spain during the Franco regime, when the inclusion of a representative of the regime and the Church was mandatory on every cooperative board (Bernal et al., 2001).

Disparities in farmers' needs for marketing were another factor affecting farmers' incentives to cooperate in each country. The UK had 367 supermar-kets by 1960, whereas Spain had only 44 stores in 1962 (Guerin, 1964). In the UK, a much bigger average farm size and a higher degree of development of the national retail sector at the time – with a greater degree of vertical integration – decreased UK farmers' need to cooperate, unlike their Spanish counterparts, who had to find a way into the market for their products. The effects of multilevel governance manifested themselves during the paradigm shift in agricultural cooperation that took place in both the UK and Spain when the countries had to get their agricultural sectors ready to join the EC, replacing their mandatory bodies with voluntary organisations in order to comply with EU competition law. The changes were even more dramatic for Spain, where, after many decades of autarchy, the country opened its borders and swiftly became a key exporter to the European market.

UK agricultural cooperation in the twenty-first century

Statistics for ACs in the UK are scarce and often out of date, a discovery that other authors have encountered too (Morales Gutierrez et al., 2005; Spear et al., 2012). Spear has pointed out how the lack of reliable data on ACs is also due to inconsistencies in industry classifications (Spear et al., 2012). Besides, the Plunkett Foundation used to collate annual statistics on ACs in the UK, but lack of funding forced it to stop this exercise in 2001. In this section, most of the statistics used were drawn from the afore-mentioned European Commission country report for the UK (Spear et al., 2012), complemented by a recent Co-operatives UK report on AC activity in 2016. The latest figures suggest that there are 434 ACs registered in the UK (Co-operatives UK, 2017), down from 621 four years earlier (Spear et al., 2012). This decrease could be due to mergers of several ACs or to a different method of collating statistics. The largest agricultural sectors are dairy, then cereals, followed by vegetable production (Spear et al., 2012). Data from 2006 show that dairy and cereal farmers were the most likely to market their produce through ACs, followed by general cropping and horti-culture, and farmers with mixed farms (EFFP, 2014). It is important to point out that small-scale farm businesses in the UK are less likely to form part of farmer-controlled businesses (FCBs); (EFFP, 2004).

As part of their work for the 2012 EC report, Spear and colleagues analysed the Eurostat Farm Structure Survey and concluded that most agricultural sectors have witnessed a decline in the number of farms, apart from the F&V sector, which has remained fairly stable and is formed by a heterogeneous mix of small and large farms. The farming crises of BSE (mad cow disease) and foot and mouth disease hit UK farming hard from the late 1990s to 2001 and highlighted the vulnerability of the sector. In 2001, the UK government set up the Curry Commission to analyse the factors that contributed to the crises and lessons learned. The final report published by the Commission was titled *Farming and food: A sustainable future* (Curry, 2002) and it informed most of the agricultural policy during the last Labour government, until 2010. Following the report's recommendations, the English Farming and Food Partnerships (EFFP) body was set up to encourage collaboration in the farming sector and work with existing ACs. However, it is worth pointing out that EFFP focused on pro-moting "farmer-controlled" and/or "farmer-owned" businesses, without link-ing to or mentioning the cooperative movement. Since the publication of the Curry report, government efforts have focused on "collaboration" rather than on "cooperatives" as such (Curry, 2002; EFFP, 2004). In addition, there was an emphasis on vertical collaborations within supply chains to compete against imports and to be better aligned with large supermarkets' supply requirements.

Food from Britain (FFB), the body that replaced the Agricultural and Horticultural Cooperation in 1983, continued to exist until 2009 (Food from Britain, no date). Described as a "quasi-governmental agency", FFB was officially presented as the vehicle for funding a revival of "regional food". However, FFB received criticisms from civil society organisations because, instead of reconnecting British farmers with British consumers, FFB was operating as an export agency, mainly focusing on fostering export activity (SpinWatch, 2005).

Currently, agricultural cooperation work is not explicitly included on DEFRA's agenda, with support being more based on words than on money or resources (Spear et al., 2012). Despite not being on the policy agenda, DEFRA commissioned a survey in 2013 to learn more about ACs and POs, with the aim to improve the competitiveness of the horticultural sector in the UK. Despite being advertised to thousands of farmers on the NFU website, only 38 respondents answered the survey (EFFP, 2013). Further details about this survey are discussed in Chapter 7.

The Curry report continued to be the basis for the Labour government's food and farming policies until 2010, when the Coalition government took power. A few months later, in October 2010, the UK government discontin-ued funding for EFFP, which became the subject of a management buyout and was privatised and renamed European Farming and Food Partnerships.

In Scotland, the SAOS has continued to provide support and development services for ACs since 1905. SAOS receives around 25 per cent of its income from government funding and the rest from members' fees and service delivery. The success of SAOS has put Scotland ahead of England and Wales in terms of cooperativism in farming (Spear et al., 2012).

In Wales, the WAOS was founded in 1922 as the central body for agricultural cooperation. WAOS has not been supported financially by the government since the mid 1980s; WAOS was funded again from 2000 to 2010, but, as in England, funding was mainly targeting "farmer-owned" ventures rather than specifically promoting ACs. After government funding ran out, WAOS was transformed into a workers' cooperative that enables voluntary and joint assurance and quality marketing, operating as an agri-food consultancy enterprise financed exclusively from fees earned (WAOS, 2011). Currently, the Wales Co-operative Centre covers, but does not specialise in, ACs; this centre is mainly funded by the European Regional Development Fund and Welsh government, with additional funding from Comic Relief and the Nationwide Foundation (WAOS, 2011).

In summer 2012, the Welsh government announced a Co-operative and Mutuals Commission that was put in place to make recommendations on how to grow and develop the cooperative and mutual economy in Wales in order to create jobs and wealth (Burge, 2012). In February 2014, the Minister for Economy, Science and Transport launched the Commission's report, which discussed and documented the contribution that cooperatives and mutuals make to the Welsh economy. The Commission concluded that agriculture, forestry and food processing are "vitally important sectors that need support to innovate and diversify in the face of significant economic pressures" (Welsh Co-operative and Mutuals Commission, 2014:34). In terms of financial support, the Commission recommended the creation of a fund for grants and loans – Co-operative and Mutual Finance Wales – dedicated to cooperative, mutual and other social enterprises, administered by existing financial intermediaries and accessed via Social Business Wales. These findings indicate a renewed interest in cooperatives in Wales (Conaty, 2015); it remains to be seen if this interest will inform Welsh Brexit policies, and, if so, what the impact on ACs might be.

In Northern Ireland, the Ulster Agricultural Organisation Society (UAOS), which was part of the Federation of Agricultural Co-operatives in Great Britain and Ireland (FAC) earlier in the twentieth century, no longer exists. Despite agriculture's relatively high contribution to Northern Ireland's economy, UAOS was denied more financial support for ACs recently, after claims pointing out they had been severely underfunded went unheard (Spear et al., 2012).

Tables 6.1 and 6.2 list the top 10 ACs in Spain and the UK. Data by turnover for the UK are dated from 2010. A more recent report from Cogeca includes other statistics from the British AC sector (excluding turnover), highlighting again the lack of reliable and centralised data available (Cogeca, 2015). More statistics for each country can be found in Appendix I. No UK or Spanish ACs are listed among the top 50 EU ACs by turnover. Only six Spanish ACs

Table 6.1 Top 10 UK agricultural cooperatives by turnover

Company	Sector	£ million
Openfield Group	Grain, inputs and marketing services	743.751
Fane Valley Cooperative	Meat and milk processing, supplies	553.888
First Milk	Dairy	460.087
Arla Foods UK	Dairy	454.263
United Dairy Farmers	Dairy	421.482
Mole Valley Farmers	Agricultural supplies	407.793
Anglia Farmers Ltd	Crop marketing and supplies	247.447
Berry Garden Growers Ltd	Soft fruit producers – marketing	212.851
Fram Farmers	Agricultural supplies and marketing	184.536
GrainCo Ltd	Agricultural supplies and storage for grain producers	165.588

Source: Adapted by author from Co-operatives UK (2016)

Table 6.2 Top 10 Spanish agricultural cooperatives by turnover

Company	Sector	€ million
Coren	Agricultural supply, processing and export	982
Grupo AN	Agricultural supply, processing and export	673
DCoop	Agricultural supply, services, export	565
Anecoop	Marketing of fruit and vegetables, services, export	539
Covap	Meat, supply, credit, services, export	373
Cobadu	Meat, milk, supply, services, shops	231
Unica group	Vegetables, processing, shops	202
Acor	Sugar, feed, petrol, services	200
Actel	Cereals, fruits, dried fruits, olive oil, agricultural supply	200
Camp D'Ivars D'Urgell	Cereals, meat, feed, services, credit, agricultural supply, shops	190

Source: Adapted by author from OSCAE (2015)

and three UK ACs made it to the top 100 in 2014 (Cogeca, 2015). For Spain, Coren was listed as number 57, Capsa as 76, followed by Group AN in position 78 in 2014, followed by DCoop (number 86) and finally Anecoop, listed as 94. The UK's First Milk appeared as number 71, Fane Valley was number 75, and United Dairy Farmers was in position 90 (Cogeca, 2015).

Spanish agricultural cooperation in the twenty-first century

As discussed in the historical section, the end of Franco's dictatorship and the adoption of democracy brought a period of transformation and opening for ACs in Spain. In the 1980s, with political devolution, legal responsibility for ACs was transferred to the newly established regional autonomous governments, a development that still exists and complicates the remit of national policies (e.g. national law might only affect supra-autonomic ACs, which are those operating in more than one of Spain's 17 "autonomous communities" with strong regional governments).

Spain was one of the last Eurozone countries to modernise its agriculture, but soon went through similar intensification, industrialisation and technological developments as UK farming had experienced decades earlier. Nowadays, Spain is a key food producer and exporter in the EU market (Bijman et al., 2012).

In the first decade of the twenty-first century, the development and consolidation of ACs continued. In 2009, the CCAE (the Spanish Confederation of Agricultural Cooperatives) changed its name and became Cooperativas Agro-alimentarias to project a modern and more businesslike image. Cooperativas Agro-alimentarias represents around 3,800 ACs integrated in 16 territorial unions and federations and Agrocantabria (a second-degree AC in the region of Cantabria; Cooperativas Agro-alimentarias, 2016).

According to Eurostat data, agriculture accounted for 2.45 per cent of Spanish GDP in 2010, and, despite a constant decrease in the number of farms in Spain (at an average of 3 per cent across sectors from 2003 to 2008), this rate is still below the 8.3 per cent EU average. This picture links with the still traditional atomised character of Spanish farms, which, as will be discussed later, is also reflected in the atomisation of the cooperative sector. Productionist discourses blame this atomisation as one of the main barriers to improving competitiveness in Spanish agricultural cooperation. Small farms mean more farmers, which in turn mean more complex heterogeneous members in ACs and reduced bargaining power when dealing with the highly concentrated retail sector.

As with the UK, the entry to the European Union shaped Spanish agriculture, farmers' incentives for cooperation and, in the case of Spain, a growing approach to exports after a period of autarchy during the Franco dictatorship years. Nowadays, according to OSCAE (the national body monitoring cooperativism in agriculture), 31 per cent of Spanish cooperatives are involved in export activities, and about 26.5 per cent of agricultural cooperatives' income comes from exports, a trend increasing year on year (OSCAE, 2015). In the 1980s, with the return to democracy and Spain's entry into the EU, the number of cooperatives increased by four. However, despite the high number of cooperatives, 27 per cent of them account for 86 per cent of cooperative trade. The remaining 73 per cent have less than €5 million activity, a low level compared with the EU average of more than €10 million euros (OSCAE, 2015). As a result, cooperatives in Spain are perceived as weaker than in the rest of the continent owing to the high number of existing cooperatives, most with very low averages in terms of number of members, paid workers and share of trade. However, cooperatives enjoy a high level of trust by Spanish farmers and are perceived as an extension of their farm, perhaps because, owing to the atomisation of the sector, they are still embedded in the fabric of local communities and social relations.

The degree of associationism – a common term used in Spain to refer to farmers coming together to cooperate – increases with the following variables: the higher the level of education, the lower the age of the farmer and the bigger the farm, the more likely the farmer is to cooperate.

In contrast to the UK, cooperative law in Spain is very complex. Spain is the European member with the most pieces of cooperative legislation (Giagnocavo and Vargas-Vasserot, 2012). The country has a double layer of legislation: a state law for cooperatives active in two or more autonomous communities and a second layer at the autonomous community level, with devolved powers and legislation for those cooperatives operating in a single region. This myriad of laws that get updated and approved at different times complicates the legislative picture enormously. The result is a weak national framework, frequently criticised for being inadequate and unable to adapt to an increasing competitive and globalised food sector. However, there are no plans for a new national, overarching legislation, as the negotiation process required would be too cumbersome (Giagnocavo and Vargas-Vasserot, 2012). The Basque Country is said to have the most advanced legislation, but the region does not have a high concentration of ACs, and so not many farmers benefit from it.

Additionally, Spanish cooperatives have a well-defined structure with three mandatory corporate bodies: a general assembly, a management board and a (weak) external regulatory body called "intervenors". The legal framework dictates the governance structure of cooperatives, with little room for change, but it allows non-members to join the management board, as well as patronage-proportional voting (Giagnocavo and Vargas-Vasserot, 2012). Historically, the influence of IOFs' models has been clear in Spanish cooperative governance (Giagnocavo and Vargas-Vasserot, 2012), but, since 1999, cooperative legislation has not been amended at the same pace, and, as an example, gender policies have not yet entered cooperative legislation.

Other factors affecting the governance of ACs are the atomised character of the sector and the existing legislation that encourages localism and excessive concentration of power by the presidents of many small and medium-sized cooperatives (Giagnocavo and Vargas-Vasserot, 2012). For this reason, the government and big players in the cooperative sector are placing great emphasis on mergers; their efforts have been translated into policy and materialised in the form of a national law. In 2012, the year the UN was celebrating the International Year of Cooperatives, the new Law for the Integration of Agricultural Cooperatives was proposed and it was approved a year later. This law was designed to encourage mergers in order to foment the consolidation of a very atomised sector. Chapters 5 and 7 explore the details and effects this new cooperative law is having on the AC sector and on farmers.

Summary of ACs in the UK and Spain in the twenty-first century

The first section of the chapter offered a historical account of the development of cooperative ideology and ACs in Europe, paying special attention to the specific cases of Spain and the UK. The second section of the chapter discussed current trends in the EU AC sector, with a focus on the F&V ACs and considering the differences and similarities in Spain and the UK. The effect of multilevel governance (MLG) has been clear in both countries over the years, as preparations for integration into the EU Common Market started a paradigm shift in national models of agricultural cooperation. For the UK, joining the EU meant the end of the marketing boards, whereas, in Spain, the sector had to open to international markets after decades of autarchy under Franco's regime. Currently, the CAP continues to encourage and recognise the value of cooperation in agriculture, but is supporting more liberalised POs over the stricter, more socially committed ICA cooperative model. The emergence and growth of POs has been discussed, providing an introduction to how POs have affected the AC sector, a topic that will be explored in more detail later in the book.

This section has also reviewed how government support and funding for ACs in both the UK and Spain have varied over the years. The AC sector went through a process of transformation when both countries joined the EU. In Spain, a clear export strategy and the need to access international markets strengthened the role of cooperatives. The number of ACs in Spain is very high compared with other EU countries (approximately 3,000 in 2012), but this is expected to gradually change after new legislation to facilitate mergers was introduced in 2013. However, there is already a strong level of market concentration: 16 per cent of Spanish cooperatives trade 75 per cent of the total trade. Mergers are seen as the solution to improve the competitiveness of the sector (OSCAE, 2015). This trend towards consolidation has also been taking place in the UK, mimicking the processing and retail sectors.

With regard to legislation, clear differences between the UK and Spain have been highlighted, reflecting the different degree of importance cooperatives represent for the economies of these two countries. In Spain, an extensive legislation specific to cooperatives has been developed over the years, whereas the UK only has a special provision under the Company Code. In most of the UK, the disjointed array of policies focusing on collaboration since the influential Curry report has only achieved a fragmented impact. The exception has been Scotland, where the convergence of policies has resulted in a strong agricultural cooperative sector. In the UK, the existence of state marketing boards until relatively recently and the early development of a highly concentrated retail market created a policy context that lowered farmers' motivation and need to form bottom–up cooperatives.

Note

1 The Commonwealth period ran from 1649 onwards, when England and, later, Ireland and Scotland were ruled as a republic following the end of the Second English Civil War and the trial and execution of Charles I.

7 Consolidation of the agricultural cooperative sector
From Farmway to Mole Valley Farm and Anecoop in the sea of plastic

From Farmway to Mole Valley Farm

Farmway was an agricultural merchant formed in 1964 from the amalgamation of three ACs, namely, Teesside Farmers, Northern Farmers and East Yorkshire Farmers. When it started, Farmway had approximately 2,000 members, and the initial investment required to join the AC was £25. It was the leading agricultural and rural retail business in the north, specialising in animal feeds, the processing and dressing of cereal seeds, fertilisers, and also the drying, storage and marketing of grain.

In 2013, while this research was being carried out, Farmway was bought by Mole Valley Farmers (MVF). When a long-standing member of Farmway who had sat on the board of the AC for many years was interviewed as part of this study, the AC he had been a member of for decades only existed in a residual form, as it was in the process of being incorporated into MVF. The acquisition of Farmway meant MVF expanded its geographical presence, and the AC now operates from Cornwall to the Scottish Borders.

MVF was founded in 1960 by a small group of farmers around South Molton who, according to the AC's website, were concerned by the discriminatory practices and the large margins being taken by many of their input suppliers. Today, MVF is the fourth largest FCB in the UK by turnover and the top supply cooperative (Oxford Farming Conference, 2014) owned by more than 8,000 farmer shareholders (Co-operatives UK, 2017).

In contrast to Farmway, MVF decided to bring in a new customer base by going into retail and selling a wider offer of products, including horse-riding gear, clothes, pet food, shooting equipment and garden and homeware, to reach beyond farmers to a wider audience of non-members.

The company has more suitably qualified persons[1] than any other agricultural company in the UK, around 350 in total. MVF has ten Mole Valley Farmers branches and three Bridgman's stores across the south west, 38 Mole Country Stores, two Cox & Robinson direct farm outlets in the south and east of England, and 11 manufacturing sites across England and Scotland (Mole Valley Farmers, 2016). MVF has a focus on precision technology and on continuing to grow and develop (Mole Valley Farmers, 2016).

MVF's family of enterprises also includes: Molecare Veterinary Services, Renewables, Farm Buildings and Mole Valley Forage Service (the last one is a joint venture with Roullier Group, a large international fertiliser manufacturing and distribution business).

In MVF's report from 2015 analysing the AC's performance in 2014, Mole Country Stores, of which the principal constituent parts are the former CWG Country Stores Limited (a chain of agricultural merchants' stores involved in the sale of agricultural raw materials, live animals, raw textile materials and semi-finished goods), and Farmway reported a positive contribution to MVF's financial results that year. However, as the report highlights, the underlying results fell short of the expectation of managing the transition of these investments into profitable growth-oriented contributors (Mole Valley Farmers, 2014). This last point exemplifies a common trend in the wider AC sector: the increasing preference for growth through mergers and acquisitions, even if expected results are not always achieved, an issue discussed in Chapter 2.

The members' perspective

This section presents the main themes emerging from the interview with a long-standing member of Farmway who was also part of the management group of this AC. The interview took place soon after it was announced MVF was buying Farmway and the acquisition process was taking place. The mains topics covered in this section are: the degeneration of the cooperative spirit, the contradiction of cooperatives' competitiveness, disembeddedness from local areas and the constraints posed by the policy context.

Degeneration of the cooperative spirit

The interviewee spoke with disappointment about how, out of approximately 2,000 members, only 30 or 40 would attend the annual general meetings (AGMs). He shared an anecdote about a driver from Farmway who was delivering a product to his farm when he was a director of the AC, but the driver did not recognise him and said to him: "the boss of this farm owns Farmway"; this episode was, for the participant, the best example that showed how the concept of Farmway being a cooperative was not strong among the membership, nor was the fact that members, not an individual, owned the cooperative. When asked what he thought Farmway's members were likely to think about the buyout by Mole Valley, he very confidently expressed his disenchantment with the cooperative form, as the following quotes reflect:

> I think that the general answer to that question is between they don't care as long as they get their money back and it's a good thing because there are products in the shops, essentially.
>
> (Farmway member)

I: The farmers are saying no, we've had enough of all this, we're going to go down the route of diversifying and selling our own meat, dairy product or whatever. But I think we'd now say that the cooperative thing is dead. I think the cooperative structure within the farming industry has probably disappeared. The company structure, the farmer-owned company structure is now much more suitable in this country.

R: For supply, or marketing?

I: For everything. It works better.

(Farmway member)

The cooperative structure works extremely well for so long and then it becomes necessary to [pause] once you start getting the squeeze, because you're really beginning to start affecting the market, then I think the cooperative structure is a weak structure. It's a very defensive structure but it's not a very aggressive structure.

(Farmway member)

I: I think the engagement of farmers with cooperatives stopped a lot of years ago, as cooperatives.

R: Can you tell me when and why you think it stopped?

I: Because they weren't delivering a benefit, a perceived benefit . . .

R: Do you think it matters? Do you think they just did their job at the time?

I: I'm disappointed because I think that they'd still be valuable, but at the moment I must admit that I'd say that I don't really think it matters. And I think, I mean, you know, industries have history of shooting themselves in the foot.

(Farmway member)

Even then, the fact that it was a cooperative was weakening in the minds of its members. The fact that it was a good agricultural merchant was the driver really, rather than it was a cooperative. I have a rule that is the father, son and grandson rule. The father sets up the cooperative; the son keeps with it because he thinks he better keep father's thing up, and the son marries in it but he starts to wander off a little bit, he doesn't buy everything and his son basically packs everything up [. . .] The grandson, yes, the grandson . . . The real thing is the grandfather is the only one that experienced the problems that were there when there wasn't a cooperative and therefore set the cooperative up and of course, the son might have done something to, you know, he might have listened to his father talking about and so on, but then the grandson, you know, things have moved on since then and go in a different way.

(Farmway member)

This final quote highlights how the lack of involvement in the original struggle that generated the motivation and level of organisation from farmers to set up any given AC means new members often are unware of those origins. By not being part of the struggle and the negotiation of the rules and the culture of the AC, members feel more like customers than owners if no education programme to keep that knowledge alive is introduced. This point is discussed in more detail in Chapter 5.

From being insignificant competitors to being unable to compete

When Farmway was growing, many supply cooperatives remained trading despite not being very competitive, or, as the interviewee put it, "they were allowed to exist" by a handful of large competitors. Cooperatives were small and not federated. Farmway was an amalgamation of three cooperative societies. They didn't work in a federal way until much later on. At the time, there were a very small number of very major suppliers in the UK, such as BNSN and Kenneth Wilson. Being part of a large trading group, they could afford to have a presence in the UK without worrying about making a profit in north-east England or considering small ACs as competitors.

Many ACs were not strong enough to be considered serious competitors; however, their added value was based on the close relationship with their members:

> The principal value was actually the problem of dealing with, for a big company dealing with a very large customer base, while the coops knew who was going to pay the bills and who wasn't with a fair degree of accuracy and they had the staff on the ground to go and talk to Fred and decide whether he was going to buy this or the other and go to his kitchen and talk to his wife or whether not to do that and [. . .] So the principal thing that the coop had was a direct connection to the customer.
>
> (Farmway member)

Over time, as average farm size got bigger, the situation changed, and ACs lost power to larger farmers. It was easier for the big companies to deal with the cooperatives. But, when farms started to get bigger, and logistics got more complicated, and lorries got bigger, then it became easier for large companies to deal directly with big farmers without going through the cooperative and the cooperative share.

At the same time as farms got bigger, margins in agriculture shrank, so that the ability of a supply cooperative to give a bonus on purchases disappeared, and, with it, the financial benefit for members decreased. Cash-strapped members did not see the logic in paying a bit extra today to get it back in dividends the following year:

> And that, that loss effectively really, is what stopped people thinking about it as a cooperative.
>
> (Farmway member)

As a result, Farmway decided not to offer bonuses on some products; the first one to be added to this list was pesticides. This caused strong disagreement among members at the AGM that year. In the 1970s, mergers, instead of federation, started to happen more quickly:

> Yep, they were just merging, they weren't federating, they just merged together and largely that was because of the financial pressure on the small ones, you know, I know Farmway was formed because basically the three it was made up of couldn't, just basically couldn't survive, the financial pressure.
>
> (Farmway member)

Policy barriers

This financial pressure came, not only from shrinking margins, but also from lack of fiscal support from the state. The government would not allow cooperative shares and interest to be classed in the tax-free category.

Disembeddedness from local area and loss of local knowledge

The long-standing Farmway member claimed buying groups such as Anglia Farmers are growing as they can demonstrate significant savings if farmers place purchases via them. Anglia Farmers, for example, accounts for one-tenth of all the UK's farm inputs (Anglia Farmers, 2016). Logistics, according to the interviewee, are another key aspect that is making traditional marketing ACs redundant. Similar to "just in time" approaches in food retail, the infrastructure to deliver agricultural goods means suppliers do not need to go through the AC's warehouse, as used to happen, and so the whole process of delivery has shrunk. These changes have altered the way the industry works. An additional factor that hinders ACs' ability to borrow funds was an increasing loss of local knowledge, as local banks grew apart from ACs and gradually lost their knowledge of how ACs in the region were working, as this quote reflects:

> Yes, the trouble is that the parameters they use now to decide whether they are going to lend you any money are not based on the ability of the company to repay but on the security of the company that exists based on some computer programme that tells them and the computers don't know how to deal with cooperatives basically, that is a simplification of the situation [. . .] there is so few of them essentially, so it's not worth the banks going out of their way to educate a number of people to deal with it and so they just don't and so West Cumberland converted themselves into a company and Farmway as well because that did the governance much easier and when Farmway came to be taken over by Mole Valley, they sold the company to Mole Valley.
>
> (Farmway member)

All these factors suggest that cooperatives are increasingly losing their financial and social embeddedness in local areas, following strategies to adapt to lower margins and savings and to new supply chains. This growing disconnection from local areas seems to be impacting on the knowledge that, not just members, but other external actors such as banks have of ACs that were previously closely linked to a geographical area.

Policy context constraints

The reductionist way of measuring ACs' performance mentioned in the previous point will be discussed in detail in Chapter 10. An interwoven theme that emerged in interviews was the application of market parameters to ACs by the Competition Commission. The Commission treats cooperatives as companies rather than as individual farmers coming together to form a group. However, even though this might sound unfair to individual members, there is also a danger in ACs getting too big. For example, Arla Food's own members demonstrated outside the cooperative factory during the milk price crisis in 2012 to complain about the low payments they were receiving (Birchall, 2012). Some farmers argue that Arla's and other large cooperatives' behaviour does not differ from private dairies, and that they are detached from their farmer-owners' control (Hanisch and Rommel, 2012).

The experts' perspective

This section presents the themes that emerged from the interviews with non-farmer experts, including two representatives from civil society organisations, one academic, one policymaker and one participant from each of the bodies that represent cooperatives in the UK: Co-operatives UK and SAOS (in Scotland).

New integrated Cooperative Act

Several participants spoke about the new integrated Cooperative Act approved in 2014, which was discussed in Chapter 2. For the Co-operatives UK representative, for example, the Act was a positive move, as it renamed the original Industrial and Providence Society legal form that "did not even have the word cooperative in it". The participant also noted that the Act was expected to bring further benefits in the future: for example, if/when the movement next campaigns for further legislation changes, there is only one Act to focus on, rather than having to think what piece of legislation is better to amend, as was the case before the integrated Act was introduced: "it's like the kind of the start rather than the end of the process". Other interviewees also had positive views on the new Act:

> [O]n consolidating the legislation on the new Act, I think that's very important because previous to this the legislation was traceable to several different Acts, so it was actually very difficult for people to follow the law

and to keep up to date. It's not just for the farmers' coops themselves, it's for the professional services, the solicitors, for instance, the lawyers who advise them, very often, they would rather simply persuade them to change to a company's Act model because it's so much easier and familiar to them than the coop law which required quite a lot of specialist knowledge to be able to trade the coop law, so having a consolidated Act will be very helpful. I think it will lead to fewer coops converting to the company format.

(SAOS rep.)

Just before the new Consolidated Cooperative Act was approved in August 2014, the withdrawable share capital limit was raised from £20,000 to £100,000 in January 2014 (Co-operatives UK, 2014b):

The main thing is the withdrawable share capital limit being raised from £20k to £100k and it doesn't get reviewed very often, the last time I think it was 16 years ago. ACs in the UK generally are under-capitalised compared to European counterparts and so this would improve this situation.

(Plunkett Foundation rep.)

However, because cooperatives are not able to buy non-withdrawable shares, it is very unlikely that the members would invest more than £20,000. What was the point of raising the investment limit then? For some interviewees, it was about making the most of this policy window in which cooperatives made it to the policy agenda:

Well, the £100k is more of a long . . . it was moved much higher because it only gets reviewed . . . it might not get reviewed for 20 years, that's the reality of coop legislation . . . even if you've got support of government, it is never at the top of anyone's list, so the opportunity was taken to go high. There were ACs calling for the £100k limit so I presume there is the basis in that there's the reality that some of their members could invest at a higher level, not necessarily £100k but more than £20k.

(Plunkett Foundation rep.)

They were one of the main sectors which was really supportive of going for such a high threshold, one of the main benefits is providing access to technologies and new production techniques and that kind of thing . . . to remain competitive in agriculture farmers need access to more and more expensive and more complex technology. I think there was an idea that in order to get that large scale investment into the agricultural coopera-tive, there needed to be a bit of a steep change in terms of the amount of money members could actually put in in order to make that happen, and not members.

(Co-operatives UK rep.)

Although this sounds like a positive development for cooperatives, the above quotes suggest that large and highly mechanised farmer members of ACs were actually lobbying for the increase to be able to fund further investments. It could be argued that this high limit might foster further consolidation in the sector by damaging those smaller cooperatives with fewer members and less financial capital. Glenn Bowen, the enterprise programme director at Wales Co-operative Centre has warned about the negative side of the limit: "we do have reservations about the size of the new limit [. . .] there is a risk members who are able to invest more may have an undue influence on the democratic structure of the business" (Glenn Bowen's quote on the Co-operatives UK's website, Co-operatives UK, 2014b). However, one of the interviewees suggested that the amount of minimum trade rather than investment is in fact the more common excluding factor for ACs' members:

> I think it is encouraging for more people to hold more capital in cooperatives. So it won't necessarily increase the scale of cooperatives but . . . and bearing in mind this is not an upper limit so people do not have to invest massive amounts of money to become a coop, as long as there is not a lower limit that excludes smaller members, then that should not be an issue. What it is often an issue of ACs is that there is a minimum of trade you need to do with a cooperative in order to be a member and that it is often the excluding factor rather than shared capital requirements.
>
> (Plunkett Foundation rep.)

UK government's influence in the sector

Although, in theory, governments are supposed to have a minimum involvement with cooperatives in order for the latter to protect their autonomy (Henrÿ, 2012), the case of ACs has been very different throughout history. For more than a century, the UK government has shaped, fostered and hindered the development of ACs in farming, depending on the political and socio-economic context at the time. The quotes below reflect these changes in government support for ACs; interviewees were asked about their views on whether the current number of ACs had gone down or up and why, and these were their responses:

> We do not have the data unfortunately, but the level, the total level of cooperatives has been fairly static since the 1930s, there has been mergers and things that take into account some, but there isn't this new growth, so I think, from the perspective of the farmer, you do not have that many options. You might be able to join an established cooperative that you might or might not feel inclined to join or you go alone. There are some that have happened in terms of quite informal cooperation but we haven't seen significant growth.
>
> (Plunkett Foundation rep.)

Another landmark was the Curry review[2] ten years ago now. It really emphasised the importance of cooperation and cooperatives in sustaining farm business, emphasising support is needed etc. When that was set up, funding did not come to us, the organisation that had been doing it for decades, it went to the English Food and Farming Partnerships.

(Plunkett Foundation rep.)

Historically, if you look at Scotland, we had many, many more cooperatives 80 or 90 years ago than we do now. Maybe three times as many. But of course there have been vast changes over the years, both in the market place and in transport, and politically. So for instance, for 50 years, from the 1930s to the 1980s, in the UK there were many statutory marketing boards that have the powers to sell or to buy all farm produce. So we had the Milk Marketing Board [. . .] and this purchased all the milk from farmers, so we did not need to have cooperatives to sell the milk at that time.

(SAOS rep.)

Before the marketing boards were created, Scotland had approximately three times more ACs; the introduction of the marketing boards reduced the need to cooperate and, thus, the number of cooperatives, but then, when the UK joined the EU, the development went full cycle:

And then when marketing boards were made illegal, when we joined the common market, when those monopolies were no longer allowed, in the last 20 or 25 years we've seen more marketing cooperatives develop and we've been involved in developing many of them.

(SAOS rep.)

Whereas Scotland still has SAOS representing the interest of ACs in the country and receiving financial support from the Scottish government, there has been a gap in England and Wales that Co-operatives UK is currently attempting to cover:

I think there is a vacuum in England that there isn't in Scotland, there is no organisation that is pulling together the agricultural coop movement in England and Wales, so there is a space to kind of grow a little bit there.

(Co-operatives UK rep.)

This representative vacuum has also been noted in the European sphere:

I: There is no equivalent of SAOS for England. Since 2010 there has been no representation for agricultural cooperatives at European level, which is bad, there's been no real voice for agricultural cooperatives in English and Welsh policy making.

R: And before 2010?

I: What used to happen is that EFFP used to fill that role when DEFRA funded and then I think they stopped in 2010.

R: Are there any clear plans to reintroduce that role?

I: I don't think so at the moment. We are trying to do some tentative steps.

(Co-operatives UK rep.)

Co-operatives UK has recently started to focus on ACs, offering a new membership package to attract new members from this key sector (Co-operatives UK, 2013a). Currently, Co-operatives UK states that there are 6,000 organisations in the country that it considers to be cooperatives (across different economic sectors, not just farming). Its representative explained how the vast majority of those 6,000 cooperatives are working men's clubs. Consumer cooperatives are the most common type of enterprise cooperative, both in terms of numbers and in terms of turnover. Also, traditional retail cooperatives, such as travel, funeral and retail societies that are independent from the Coop group but operate in a similar way, are very common. Why then is Co-operatives UK showing this new interest in ACs now?

I guess a good number of the largest coops in this country are agricultural coops, so if we're there to campaign on behalf of coops it would be strange to at least not acknowledge the fact that some of the largest ones are agricultural cooperatives.

(Co-operatives UK rep.)

When asked if Co-operatives UK would still be adding to its books ACs that might not necessarily comply with the ICA principles, the interviewee offered the following response:

[I]f some of the largest, more high profile coops are not behaving very cooperatively, then that represents a risk to the idea of public understanding of cooperatives in general, so we'd want to try and encourage them to be more cooperative and we can only do that if we have a relationship with them.

(Co-operatives UK rep.)

Lack of funding to support or monitoring the sector

The Plunkett Foundation had been collecting statistics on the AC sector nationally and internationally for nearly a century when, in 2001, funding for this activity came to an end. The participant was asked whether the Curry review from 2002 had been related to this change:

Not necessarily. I think the NFU had a large . . . the NFU wanted a new independent body set up rather than an organisation that, you know, we had some baggage, we had been going on for 75 years, so . . .

(Plunkett Foundation rep.)

This reference to the historical conflicts between the NFU and the AC movement in the UK is in line with the findings from the literature review (Knapp, 1965). When asked why the NFU might have wanted an independent body, the participant answered:

> We don't know [in louder voice]. There'd been certain disagreements with the NFU over the years about cooperatives, so, but when there was the ambition to set up the Agricultural Organisation Society, it was not something that the NFU supported because they thought it would be a secondary voice for farmers and it would dilute the voice of farmers and they wanted one united voice. Really I think it comes back to that really. We were seeing as an organisation that promoted cooperatives, we had a certain view of the world, a certain view of why cooperatives are needed, and why they are important and not everyone agrees with it, so the EFFP and the emphasis on collaboration it meant that they had broader scope to do things slightly different. Lots of the focus under EFFP was supply chain collaboration. People have called it vertical cooperation, this is a term that seemed to crop up in the last couple of years, which is not really cooperation at all.
>
> (Plunkett Foundation rep.)

Since the privatisation of EFFP at the beginning of the Conservative and Liberal Democrats Coalition in the second half of 2010, there has been a vacuum in the monitoring of ACs. The Plunkett Foundation stopped its annual reporting on ACs following a severe reduction in available funding at the end of the 1990s. The brief UK country report included in the European Commission study from 2012, quoted several times in this book, is a reflection of the lack of available data on ACs (Spear et al., 2012). The lack of funding was also quoted by Co-operatives UK as the reason why, when the large cooperative Dairy Farmers' of Britain collapsed in June 2009, DEFRA was asked to provide more policy direction on the governance of ACs, but the new Coalition government moved funding away from this work and also privatised EFFP. Co-operatives UK is now trying to cover this monitoring gap.

Changes in government support were also happening while this book was being written:

> If you look at the abolition of some countryside and rural bodies, whether it is the Commission of Rural Communities and currently this discussion going on as we speak now about ACRE, the Association of Community Rural Enterprises, they have a similar role in monitoring what is going on in rural areas . . . Why hasn't it . . . part might have to do with monitoring, but there is a minister in the government who covers, I think his title is food, farming and the marine environment, into which farming in all its guises is included, including the cooperative elements of it would fall too, but it is only one small part of it.
>
> (UK policymaker)

There has to be a focus from any government on formal cooperative structures within farming, not forcing them on farmers, but finding a way in which government can support the establishment of cooperatives in farming.

(UK policymaker)

It is interesting to mention at this point that the UK has a political Co-operative Party that has been a sister party of the Labour Party since 1927. Today, there are 32 Labour and Cooperative MPs (Co-operative Party, 2014).

The quickest route to do something through parliament, is always through the coop party [. . .] but obviously they can only do things with Labour [. . .] the good thing of having a relationship with us is that we can also reach out to Liberal Dems and Conservatives in a way that they can't do themselves, so we can create a more rounded joined-up approach; yes, it's historically, especially that we have a Co-op party that it is tied to another political party.

(Co-operatives UK rep.)

It seems a paradox that it was when Labour was in power that there was a keen promotion of vertical cooperation in the supply chain rather than horizontal cooperation in farming. When asking Co-operatives UK about trade among cooperatives, these views were expressed:

From our point of view, cooperation amongst cooperatives is not necessarily trade [. . .] A Cooperative Development Fund, like in Italy or Spain, has been put forward in Wales [. . .] In this country there has been more work on coordinating at the political level, with the Co-op Party, the Co-op group financially supporting the Labour Party, or the formation of Co-ops UK . . . It's also due to the economic culture of the UK, always trying to maintain a liberal economy with less restrictive economic models.

(Co-operatives UK rep.)

Cooperatives, it seems, while having a clear political past and still strong political present, are being integrated into neo–liberal economies: they are not vehicles for furthering social and environmental change. They are very much vehicles for ensuring those independent farmers remain economically competitive and viable.

(Co-operatives UK rep.)

I: I get the impression that farmers especially in this country, especially in England, are from a kind of social and political demographic if you like which isn't going to see cooperation as necessarily a plus, but as something of the left, and they're not interested, it's not for them, so if these businesses or cooperatives want to attract membership, then calling themselves, coming and saying: "come and join our farmers' cooperative comrade", is not going to win them their vote, or win them their memberships, they know their target audience.

R: Do you think the coop legal form is still politically charged in that way?

I: We try to depoliticise to an extent as much as possible, but it is.

(Co-operatives UK rep.)

Mergers and acquisitions trend and POs

Several participants spoke of the pressure on ACs to get bigger in order to benefit from economies of scale and protect themselves from EU competition. The quote below summarises most of the comments made in relation to this theme:

> The trend on coops has been mergers, acquisitions, consolidations. [. . .] And we have seen some failures, like Dairy Farms of Britain, that was quite a spectacular failure. [. . .] Another important influence has been POs, the status of POs, [. . .] they want you to be of a particular scale but [. . .] they don't want you to have a too high share of the market, which is difficult for cooperatives. It is natural for some cooperatives to merge because they are doing a similar things but then it is difficult for new cooperatives to form because there is that massive incentive for POs [. . .] so what we have seen is very few new cooperatives formed over that period.
>
> (Plunkett Foundation rep.)

This first section of the chapter has discussed the case of Farmway from the perspective of a long-standing member and, from there, related trends in the UK AC sector from other experts from civil society organisations. The key themes that have emerged are the degeneration of the cooperative spirit, the trend and impact of growth for ACs and the loss of farmer members' power associated with the latter. Other issues discussed were the increasing disembeddedness from local areas and loss of local knowledge of the activities of cooperatives. Interviewees also brought up points about the new integrated Cooperative Act and the effect of other government policies that have hindered or promoted ACs over the years. An interesting issue raised by participants that highlights the lack of attention paid to ACs by different UK governments is how funding to monitor ACs' activity was discontinued in 2001, making it very difficult to accurately assess current statistics and growth trends in the sector.

Anecoop in the sea of plastic

In Spain, agriculture is seen as a key mechanism for rural development and regional economies, reinforced by European policies (Bijman et al., 2012). There is a long cooperative tradition in the Mediterranean country, which has the second largest cooperative economy in Europe after Italy (Bijman et al., 2012). Spain has approximately 4,000 ACs. As will be discussed next, for many informants and policymakers, this figure is a burden and a cause of embarrassment rather than pride, as the turnover of most ACs is very low, and, thus, most of them are not considered competitive enough. Additionally, a high number of small ACs is deemed to result in inefficient expenditure from the public purse when subsidies are administered. For other participants, efforts and policies aiming to reduce this number are threatening diversity in the AC sector, but also, more importantly, diversity in farming and the rural world.

This section starts with the presentation of a case study: Anecoop, a leading F&V AC with operations and members all over Spain. A discussion is then offered introducing the views from other expert informants on Anecoop and on the new Spanish law for the integration of ACs, enacted in 2013.

Case study: Anecoop

Anecoop is the largest AC exporter of citrus fruits in the world and the second largest wholesaler, exporting about 80–90 per cent of its production, mainly to the EU (Anecoop, 2015). Anecoop is a second-degree[3] AC that has 71 associated cooperatives, across the whole of the country. Its website is translated into English, French and Danish. When asked how many different crops the AC commercialises, one of the representatives said, laughing, "I don't know, I can't remember, everything that grows on soil", which is pretty much correct judging by the catalogue on their website: all sorts of fruits, vegetables and wine (Anecoop, 2013).

The cooperative celebrated its fortieth anniversary in 2015. Anecoop was founded in 1975 by a group of 31 citric cooperatives from Valencia to gain negotiating power to deal with large retailers and to be able to deliver big contracts with other countries that no individual cooperative could access owing to geographical and size limitations (Planells Orti and Mir Piqueras, 2004). The cooperative exports its produce to 69 countries and was an early adopter of internationalisation strategies; its first subsidiary company to carry out marketing operations abroad was set up in 1978–9 in Perpignan, France. More subsidiary companies followed gradually: IFS in France, Fruchtpartner and Mercato e Ifta in Germany, 4 Fruit Company in Holland, Fesa UK in England, Anecoop Praha in the Czech Republic and Anecoop Polska in Poland. More recent ones were established in Ireland, Russia, Belgium, Denmark and China (Planells Orti and Mir Piqueras, 2004).

In 2002, a small group of cooperative members pushed for the formation of the Anecoop Corporate Group, which became formalised in 2003, when the group started to investigate how to merge all Anecoop's AC members into one large AC. Many did not agree with the idea, and, given the complexity involved in mergers, an alternative option was explored; this alternative consisted of a structured attempt to homogenise the management of the member cooperatives while maintaining their independence, but "trying to change a hard reality that takes us nowhere" (director and deputy director of Anecoop, in an article written by them; Planells Orti and Mir Piqueras, 2004:402).

The 31 original cooperatives became 109 in 2004. In 2015, Anecoop had 71 cooperatives. Some of them merged into larger cooperatives; some of them were allowed to go out of business (Planells Orti and Mir Piqueras, 2004). Despite the reduction in its number of cooperative members, volume and trade have continued to grow. That same year, 70,193 people formed Anecoop's team, of whom 28,206 were farmer members, 22,300 were members in other types of service, and 19,687 were workers.

Members are required to trade a minimum of 40 per cent of their produce with Anecoop, although many trade more than that figure. The destination of the remaining percentage is negotiated with the cooperative to avoid sending it to the same market or buyer. Anecoop defines its strategy as based on permanent adaptation and innovation, quality, efficiency and growth (Anecoop, 2016).

The data presented in this section focus on Anecoop in Almeria. This branch of Anecoop was chosen for its geographical location in one of the regions of Spain – Almeria – where farming is highly embedded in the industrial paradigm and global international markets. Almeria is an Andalusian province known for its miles and miles of plastic greenhouses along the coast. Almeria had a golden age in the late nineteenth century in part thanks to export of the Ohanes grape, a traditional variety cultivated in the north of the province (McNeill, 2003); the Ohanes grape has now been replaced by seedless, fatter and thus more marketable varieties, although many think that while the shape might have been improved, the taste has been lost. The explosion of greenhouses began in the 1960s. Before then, the coastal area in the southern part of Almeria did not have a significant pre-existing cooperative culture or significant commercial agricultural activity (Giagnocavo, 2012). During the 1960s and early 1970s, underground water found under some 30,000 hectares gave the area the status of a designated zone of national interest; soon, Franco's regime introduced wells and basic pumps in an attempt to increase agricultural production (Giagnocavo, 2012).

The first greenhouse was built in 1961, and soon the development of experimental greenhouses at the beginning of that decade started a gradual influx of farmers descending from the Alpujarras region in the north of Almeria (Sánchez Escolano, 2013). Giagnocavo has described the newly settled farmers' strategies to confront the dry and unfertile soil, saline water and strong winds. These farmers introduced a "technological innovation", which consisted in "putting down a layer of fertiliser, then covering this with a layer of sand, in order to keep the roots moist and filter the salt" (Giagnocavo, 2012:3). Later on, plastic and wooden posts from a decaying table grape industry were used to build the first rudimentary greenhouses that radically transformed the region (Sánchez Escolano, 2013). By 1976, 3,081 hectares were covered in plastic; this area grew to 9,657 hectares by 1984 and doubled to 18,694 in 1999 (Sánchez Escolano, 2013). The figures from 2015 reveal an ongoing expansion, with the area dedicated to intensive greenhouse cultivation reaching 29,596 hectares

(Europa Press, 2015). The area is commonly known as the "sea of plastic" and it is one of the few human-made constructions visible from space (Europa Press, 2015). Figures 7.1 and 7.2 show the extent of the transformation of this part of the Almerian region. One of the participants explained how the new methods created some generational conflict when talking about the visual and economic transformation and the growth of greenhouse agriculture in the Almerian coast:

> In the year 1960 it was zero, zero hectares and zero farmers. Well, there were the traditional ones from Berja, Dalias, Valle de Andarax, who were grape farmers; an interesting situation emerged, Serafin Mateo wrote about it very nicely, the fight between the traditional grape growers parents and grandparents and the children who were innovative.
>
> (Anecoop rep. 1)

The interviewee recalled how many young farmers came down from inland towns and villages to meet friends and started working with them as labourers in the greenhouses, but then borrowed money from the bank to set up their own greenhouses. This was followed by a growing demand for agronomists to improve issues of waste and run-offs, fertiliser quality, and so on.

The second reason for the choice of the case study that is closely related to the first reason (i.e. size and commercial success), is that Anecoop shares an experimental farm with the University of Almeria (UoA), where new trials are tested for the market. This collaboration is examined in this chapter. The experimental farm is called Fundacion Finca Experimental Universidad

Figure 7.1 Satellite view of the "Sea of plastic" in Almeria, Spain.
Source: NASA (Image labelled for reuse).

Figure 7.2 Aerial view of surface dedicated to greenhouses in Almeria, Spain.
Source: ANE (Image labelled for reuse).

de Almeria Anecoop (University of Almeria–Anecoop Experimental Farm Foundation) and will be referred to in the text as "the experimental farm" or FFE. Although there are a few other experimental farms, this one is by far the largest in Spain.

The members' perspective

This section presents the themes that emerged from the interviews with member workers of Anecoop. One of the key themes was around how to keep such a large number of members involved and informed of the latest policies and developments. When asked how/if knowledge transfer happens in the region, one of the Anecoop representatives answered: "Almeria is restless and very curious". A reflection of this attitude is that more than 80 per cent of the fruit and vegetables produced in the region are consumed in Europe, with Germany and France heading the list of buyers. When asked if this factor was the motivation for speed and innovation in farming in Almeria, one of the participants reflected:

> The history of Almeria has been a race [. . .] everybody has had to adapt to an intensive production, move the crop from natural conditions to under plastic, and this caused an acceleration of the process and I think all of

us involved in this sector we have had to follow that rhythm a bit, that rhythm that creates a dynamic that looks normal but when you speak to an olive grower or a wheat grower [laughs], it breaks all your conceptions, or he is going very slowly or we are going too quickly.

(Anecoop rep. 1)

This ongoing race to increase the intensity of what already are intensive farming methods is expensive, and not all farmers can afford it. Currently, approximately 30 per cent of greenhouses have obsolete designs, as owners refuse or are unable to invest on their upgrade (Giagnocavo and Vargas-Vasserot, 2012). The race is also having an impact on ACs, as, when farmers can afford any investments, these normally go towards paying for farm assets rather than towards equity in their cooperatives (Giagnocavo and Vargas-Vasserot, 2012).

When asked if they believed the "theory of the grandfather, the son and the grandson" proposed by the English farmer member from Farmway, which suggests the grandchildren of original cooperators do not realise the value of their cooperatives and the struggles they emerged from, one of the participants shared his views on the social networks that he believes foster stability in a cooperative:

There is a curious phenomenon; I think what gives a generational stability to cooperatives, is merging different relatives, first, second and third removed and neighbourhood relations. If I can achieve that in a cooperative every one of those groups formed by geographical areas, I mean by neighbourhoods, and by family units and every one of those groups has their representation in the board of directors, then it is a social and economically stable cooperative. Having a trusted person [. . .] so that the next day of the board making a decision that can affect me, I can have the information directly (me as a producer member), in an informal manner, that's how it works here in Almeria, a phone call, a WhatsApp message . . . once the cooperative reaches that stability, then it can enter a generational period.

(Anecoop rep. 1)

This reflection on the social aspect of ACs seems like a contradiction when compared with the ongoing growth of faceless cooperatives. However nice the social aspects sound, ACs' clients seem to be looking for something different:

[Q]uality of the produce, service and price are the three pillars considered by clients. [. . .] If they all give good service, nowadays we all give good service, and if they have quality, nowadays we all work for quality, then price is the decisive element.

(Anecoop rep. 2)

Anecoop is a second-degree AC, and so the new law fostering mergers in the AC sector is not going to affect it directly, but it will affect its member cooperatives. One of the participants I spoke to shared their views on the matter with regard to the concerns some have raised about how the law will concentrate the sector excessively and farmer members will lose control and power:

> I think what the cooperative law is fostering is integration to avoid the atomi-sation that we currently have; the thing is of course, the cooperative law frames it within the concept of integration. Integration is absorption, is to manage a company using the same parameters. In my view, I think there are other things to acquire dimension . . . in my view an adaptation period is going to be needed and a process to allow that cooperatives actually meet the require-ments specified in the integration law . . . we are working towards them . . . it's not about merging, in our case it would be about integrating, integrating the business management into Anecoop and that's what we are trying to do.
>
> (Anecoop rep. 2)

A couple of the other participants external to Anecoop had mentioned in their interviews that the AC had been losing members recently. This was an inter-esting point not mentioned by the interviewees from Anecoop until they were directly asked about it. This was their answer:

> Well, the majority of members we have lost is because they haven't integrated themselves, because there have been cooperatives that have absorbed and integrated each other and there are others, logically due to their dimension, that not having integrated themselves in first-degree structures have disappeared for lack of viability, but we are continuing to do more volume with fewer members.
>
> (Anecoop rep. 2)

This is a trend that the participants saw as ongoing, as the next quote suggests. Building on this topic, the participants were asked: Where do you see Anecoop in 5 years' time? The following is an extract of the interview that covers the participant's answer to this question:

> I: In 5 years . . . [thinking] I still won't be retired . . . [laughs]. Let's see, we have an integration project, that although it does have a certain parallel-ism with the cooperative law, what tries is, well, to increasingly integrate members in the structure of Anecoop, integrate the business management, so, where do we see it? I think it's a simple survival logical way of looking at it, we have to continue growing, we have to continue growing.
>
> R: In trade or members?
>
> I: In everything. It would not worry us not to grow our members as long as our members' dimension continued to grow. We would not mind in 5 years' time to have instead of 72 cooperatives, have 50 but that those 50 made more volume than the 72.
>
> R: So it's growth in terms of volume?
>
> I: Yes.
>
> (Anecoop rep. 1)

The above extract reflects how this cooperative's focus is placed on continuing to increase its turnover rather than number of members. In fact, the fewer but larger and more "competitive" members, the easier to agree on decisions and implement consistent standards. The next section discusses another strategy Anecoop uses to increase its competitiveness in the area of product development and market research.

University of Almeria–Anecoop Experimental Farm Foundation (FFE)

The University of Almeria bought land in 2002–3 with the plan to start an experimental farm for crop research. In the year 2003, the building works began, and activities started in 2004. The farm has approximately 14–15 hectares, of which only 10 are fenced. To put the size of this farm in context, it is interesting to note that the average farm size in the country is 24 hectares, but with great differences among regions. Currently, more than half of the agricultural holdings in Spain have less than five hectares of utilised agricultural area (UAA)[4] and occupy less than 5 per cent of the total Spanish UAA (Eurostat, 2012). The University of Almeria has a Department of Agronomy, and students carry out visits and research projects on the farm. As the representative from the FFE mentioned, the university thought it would be "logical" to reach an agreement with a "powerful" F&V business and it realised Anecoop was the biggest exporter in the sector. During the interviews, one of Anecoop's representatives mentioned that the agreement was in fact reached because one of the directors of the cooperative knew the academic from UoA who was in charge at the time. Apparently, Anecoop had already been looking for land to carry out trials:

> The collaboration took place because one of the directors of Anecoop knew one of the academics personally. A meeting was organised at which the collaboration was agreed. Anecoop had already been looking for an experimental farm for a while.
>
> (Anecoop rep. 1)

Anecoop makes use of a section of the farm for commercial purposes. This experimental farm is in fact the largest experimental farm in the country. Anecoop has a seed bank with seeds that come from either members or partners. Sometimes the cooperative also receives free seeds from large players and has several crop development and improvement projects taking place in conjunction with Syngenta:

> [W]e collaborate. In fact yesterday a colleague from Syngenta came here. Phytosanitary companies, if they commercialise a new product with high efficiency that can bring improvements, we normally test them in collaboration with them.
>
> (Anecoop rep. 2)

When asked if they are sold these new products for trials, this was the answer received:

> In principle the relationship is one of "try it and let me know what you think". They are also interested in having access to see the results with somebody neutral.
>
> (Anecoop rep. 2)

As the participant pointed to a specific place when talking about the colleague from Syngenta, a follow-up question enquiring if their visits were very regular was posed:

> Yes. You need to take into account the tests we do with seed companies for example, each company knows about theirs but don't know about the others' because there is a high degree of confidentiality, you need to realise they are competitors.
>
> (Anecoop rep. 1)

I then asked if they worked with big seed companies, to which one of them proudly answered:

> We work with the most important companies in the sector.
>
> (Anecoop rep. 2)

> With Monsanto we worked last year, and this year, I'm not sure, let me check . . .
>
> (Anecoop rep. 1)

After reviewing the FFE's annual reports, Anecoop's relationship with Monsanto and Syngenta seems to be in fact a regular one, as all the reports checked included trials using varieties and seeds from these companies (Fundacion Finca Experimental, 2013).

Some of the most recent experiments included a trial to grow tomatoes with desalinated water instead of fresh water (Chile Alimentos, 2015); water is a key resource in farming, but especially when growing thirsty crops such as tomatoes, cucumbers, and so on, in a region such as Almeria, well known for its desert climate. The farm also holds a seed bank of around 2,000 varieties of vegetable seeds native to Almeria. Farmers from across the region were visited by Anecoop's representatives who collected samples of all the varieties they could find. Many varieties were currently cultivated only by farmers over 75 years old (Foco Sur, 2011). Apart from seeds, I wanted to find out more about the free inputs Anecoop's experimental farm gets from these companies:

R: But are you given the phytosanitary products or just the seeds?
I: We have to be open. [laughs]
R: To anything, right?

I: My colleague is completing the record, the seed company ones, that's why I don't want to tell you the wrong thing. [. . .] the idea would be to work with as many seed companies as possible, it's in our benefit to be considered as collaborators; we've had some new Chinese companies that have come and we have also talked to them. And the same with phytosanitary products.

(Anecoop rep. 2)

In the experimental farm, Anecoop does not undertake basic research but experimentation and improvement of new varieties. It checks if those seeds have the properties the manufacturer claims, and, if it is happy with the results, the seeds are transferred to its cooperative members through demonstrations to establish the virtues and problems of the new variety, hybrid or crop. If, in the end, the new plant is considered suitable, one or two years later, members are officially informed that they can cultivate it and commercialise it through Anecoop. The cooperative regularly invites commercial buyers to the farm to check their experiments and new varieties:

It's simply to tell them: listen, you have to come and see my efforts to develop new alternatives, let me show you the results I'm getting, so you participate with me in those results and I'm very interested to know your opinion, because on the basis of your feedback, I will do a higher or lower level of transfer to my cooperatives. [. . .] we can set up a meeting to discuss a new future product line, we talk about metres, we talk about prices, we talk about the first step to introduce new things. We have had Carrefour [. . .] BAMA that is today a heavyweight in Norwegian retailing. Or Edeka, the top German supermarket; Carrefour is coming to the see the farm next Friday. So then, to all of those people, it's very easy to communicate what we are doing here, and if a client says: "listen, I liked that tomato, I'm interested", then the following year we do a bigger trial and we give them samples. You need to take into account that at the end of the day, supermarkets are the ones that transform our work into money, so we need to see what it is they need. What it is they ask us for, what it is they demand from us in order to harmonise it with the farmer's need to produce.

(Anecoop rep. 2)

The experts' perspective

This section presents the themes from interview data related to non-farmer experts, including two representatives from civil society organisations, one academic, one policymaker and one person from Cooperativas Agro-alimentarias, the national body that represents ACs in Spain.

Loss of members' control

When discussing the impact of mergers and internationalisation on farmer members, participants talked about changes that indicated a gradual process of disembeddedness of ACs from members and local areas. After takeovers and mergers, although the cooperative might still remain in the local area collecting farmers' produce, members often suffer a loss in their decision-making power:

> Ok, Anecoop members are not able to make many decisions, that's true, but their product is being commercialised and doesn't end up sitting there for a year.
>
> (Cooperativas Agro-alimentarias rep.)

> Of course. The member has less decision-making power? Yes, yes. Because obviously he will be given certain requirements and he will be told in order to export to US you have to comply. Or to export to Japan.
>
> (Cooperativas Agro-alimentarias rep.)

> It's true, even people very embedded in classic cooperativism models have the same critical reflection [. . .] "it's gone out of hand", "we no longer have control over cooperatives". They don't know well why, but they criticise them because they're in the hands of directives, that cooperatives have grown too much, that cooperatives only have their eyes on large exporting targets, that cooperatives have been the entry door for the green revolution, that they have been the entry door for GM.
>
> (Soberania Alimentaria rep.)

The above quotes show how farmers who are part of large ACs have to accept rather than make decisions as members. These decisions are related, not only to marketing and to which customers/countries to sell, but also to their daily growing activities, the methods and varieties they use.

The impact of POs[5]

The representative from Cooperativas Agro-alimentarias talked about how many ACs are actually registered as POs, as the volumes required to get funding are much lower. The requirements are approved in Brussels, starting with a number for minimum volume of produce that keeps being negotiated down and down until "when you realise, anyone can register as a PO" (Cooperativas Agro-alimentarias rep.).

> So for example, farmers associated with Anecoop are both members of an AC and a PO at the same time [but it is the same organisation]. ACs use the funding they receive from the PO stream to manage the market

operations or harvest and logistic costs. They present a business plan and if it approved, they receive subsidies as POs.

(Cooperativas Agro-alimentarias rep.)

When it was pointed out that there was a contradiction between the Spanish government's efforts to foster ACs' mergers and the ongoing reduction in the minimum volume required for POs to qualify for funding, the policymaker agreed:

You are completely right. I mean, that is a legacy of 10 member estates from Eastern Europe. These POs in the West are quite big, well, not as big as in Holland or Denmark but we don't do too badly. [. . .] What is a contradiction is that our development programme has funding for both [. . .] to receive subsidies based on this article you have to be an SME, and for me [the government] on the other hand, if it's an SME, it does not reach the minimum established in the Royal Decree that we have specified to be considered macro, to be fat. So we are going to give funding to the small ones and the big ones.

(Policymaker)

The thing is, POs are not required to be cooperatives, so what happens? We have other lobbies that defend the opposite to us. We are totally against POs being formed by four friends, we never decided that and we never defended it, but, we did not achieve our objective. What we have defended is that POs should be as strong as cooperatives, but what happens? If a union says the opposite and says no, instead of requesting 200,000 kilos with 2,500 is enough and then another union says the same and the exporters say the same, you know, in the end the Ministry decided what it wanted, and obviously they did not go with our view.

(Cooperativas Agro-alimentarias rep.)

The above comments reveal the complex and contradictory subsidy system in place for ACs and POs in Spain, currently being financially rewarded for being really big or for being really small. This indicates an unfavourable policy context for medium-sized cooperatives. Some ACs receive more funding also by being registered as POs:

Yes, it's EU funding. The CMO for fruit and vegetables accepts the PO form (whether a cooperative or not, they put them in the same bag) and the volumes they require are very small.

(Cooperativas Agro-alimentarias rep.)

The participant went on to explain the case of a PO in Murcia, in southeastern Spain, where a large exporter had selected seven small farmers from which he regularly buys produce and registered them as a PO to access and keep the funding:

We are against this, because that PO is managed by the exporter and they are the ones who keep the money, it does not benefit the farmers and we have reported it in Brussels.

(Cooperativas Agro-alimentarias rep.)

Impact of policy transfer

Several Spanish representatives talked about the Dutch or Danish models as examples of best practice and as an aspiration for the Spanish AC sector:

Here what we try to do is that ACs become stronger because distribution is highly concentrated [. . .] the model to follow is the Dutch or the Danish, with strong cooperatives that can negotiate prices with the distribution.

(Cooperativas Agro-alimentarias rep.)

The good thing that Europe has brought is that it's made us enter the EU market, but we haven't learnt, because if you have read the Law 13 of 2013,[6] the preamble of the law says that a single Danish cooperative is bigger than all the Spanish cooperatives together.[7] We are doing something wrong.

(Policymaker)

The Dutch and Danish AC sectors are highly concentrated. As discussed in Chapter 2, most statistics at EU level combine both AC and PO data to report on the size of agricultural cooperation in the Common Market. In the Netherlands, the market share of POs marketing fruit and vegetables is actually higher than that of formal cooperatives (Bijman et al., 2010).

New cooperative legislation and related policies for ACs in Spain

This theme had two strong interwoven subthemes that will be considered jointly here: consolidation and internationalisation.

Consolidation

Related to the impact of policy transfer and the widespread perception of Dutch and Danish ACs being the best role models to follow, the Spanish government and the national body representing ACs are directing their policy efforts towards mechanisms that aim to reduce the atomisation of the sector and achieve consolidation. In 2013, the Spanish government published the Law for the Integration of Cooperatives and other Agri-food Associative Entities. Apart from mergers, the new law was expected to create a better alignment with large retailers and reduce the current administrative burden of processing subsidy claims for so many ACs:

To correct the disadvantages produced by the aforementioned atomisation, the Government has set as a priority the development and promotion of cooperative and associated integration, in the belief that it will favour competitiveness, redimensioning, modernisation and the internationalisation of such entities, in the framework of structural measures to improve the economy and the competitiveness of the country.

(BOE, 2013)

This way of demonising people like Mercadona,[8] it's not on. Mercadona is not Darth Vader. Mercadona does its job and does it really well. And it buys and sells really well. And we, here in the first link of the chain we do it badly. So if these people [ACs] get stronger through measures such as mergers, perhaps we'll gain negotiating power.

(Policymaker)

The CAP constitutes approximately 30 per cent of the agrarian income in Spain, but then I need to improve their market that is the 70 per cent of their agrarian income, and this is not dosh, these are structural measures [. . .] from offering education – Ministry of Education – to improve roads so that the logistic facilities are better – Ministry of Public Works and Transport – to improve their ability to be competitive. We have structured it in three pillars: internationalisation, food chain and the improvement of associationism that is the Law 13 from 2013.

(Policymaker)

What I can't have is 900,000 CAP beneficiaries in Spain, of which actual active farmers are 300,000. It's doesn't work. I mean, it is what it is, I will have to live with it. But it is not really the objective of a competitive agriculture.

(Policymaker)

Here, the interviewee was referring to their dismay at having to deal with nearly 100,000 subsidy recipients, of which two-thirds are not full-time farmers. The administrative costs are very high and, thus, considered a burden and a policy problem that needs to be solved. The actual law did not initially establish a budget, but it announced the plan to develop it in order to offer financial support to reward the creation of a new cooperative form introduced by the law, the "priority agri-food associative entity" that is expected to bring about:

in some cases, the disappearance of original entities as they become integrated in a new entity, in other cases, the disappearance by takeover, the accreditation of an existing entity, or the creation of a new entity maintaining the base original entities that constitute a higher level entity that controls, at least the tasks related to the joint commercialisation of the production of all the members from the original entities.

(BOE, 2013:56583)

Anecoop seems to be already aligned to this growth strategy promoted by the law, as the aforementioned quote from one of the representatives expressed the cooperative's strategic desire to continue growing in terms of turnover, regardless of the number of farmer members or ACs is has on its books.

On the one hand, government policies are encouraging large cooperatives to grow, while, at the other end of the sector, smaller ACs are not supported to grow and not compensated by the loss of subsidies they suffer when reaching a certain size. For example, when cooperatives grow out of the size criteria to be considered SMEs, they lose 50 per cent of EU rural development funds. Competition laws are also contradictory, because, at the end of the day, ACs, regardless of their size, are still made up of individual businesses. Some authors argue this might be owing to the lack of understanding of cooperatives' needs, as policymakers do not seem to engage the sector in policy discussions (Giagnocavo and Vargas-Vasserot, 2012).

Internationalisation and embeddedness in industrial and globalised food systems

With regard to internationalisation, the Spanish government's aim is to go beyond just exporting produce and to move instead into a more strategic type of growth abroad:

> well, incrementing exports or rather incrementing the positioning of enterprises abroad, which is more than internationalising, more than exporting. The first part for you to be able to access a foreign market is to take the produce out, for Anecoop, to take the fruit out. But maybe, they can sort out instead a cooperative with a strategic member in Belarus, to be able to enter Russia. [. . .] Or see where it makes sense to export and where not. And see where you are failing and where not. In the oil sector the same thing is happening. In fruit and vegetables, there is a lot going on around internationalisation that is more than exporting. See what administrative and bureaucratic loops you are asked to jump to put a lorry in a specific market, etc. there are more things that can be done, yes, yes.
>
> (Policymaker)

> [I]t is necessary to implement measures that foster the integration and strengthening of commercialising groups of associative and cooperative character that have operations and a remit beyond that of one autonomous community; and able to operate in the whole food chain, both in national and international markets and that contribute to the improvement of rural income and the consolidation of an agrifood industrial fabric in our rural areas.
>
> (BOE, 2013)

> The export activity of Spanish ACs is growing every year, both in terms of numbers of ACs involved in exporting and in terms of volume exported. In 2013, 31% of ACs were exporters. Turnover from exports grew this same year by 13.8%, accounting for 26.5% of the whole AC sector turnover.
>
> (OSCAE, 2014)

When the policymaker was asked what they thought were the risks of an increasing volume of produce leaving Spain to be sold in other countries at higher prices, they answered: "that's the rules of the market, what can we do?". The same rules of the market also normally favour size and economies of scale, as they help ACs to survive financially in the competitive global agri-food sector. The effects these rules and trends are having on ACs and their members are discussed in the next section.

A battle against middle-sized ACs

Although the farmer members who took place in the study were not aware of the new cooperative law from 2013, the experts consulted had their own views about it. The volumes promoted by this law are quite high, which means it is targeting medium to large ACs, as, even if small ACs merged, they would not reach the minimum stated:

> I have a very small spectrum, roughly 10 cooperatives: Anecoop, Conen, Guisona, Decoop, Baco, which are going to merge next week . . . ok, say 20. Then at the other end I have 4,000. This is really, really sad. Here I have 2,000 that are garage cooperatives,[9] subsistence ones, of "the virgin such and such" [. . .] and the ones in the *middleeee*, the ones in the *middleeee* [in an annoyed voice] are the ones I want to catch.
>
> (Policymaker)

The frustration on the interviewee's voice when referring to middle-sized ACs was clear. The interviewee was drawing circles around the rectangle used to depict middle-sized cooperatives in a diagram they had drawn while speaking, as shown in Figure 5.1 in Chapter 5. The representative from Cooperativas Agro-alimentarias was also asked to expand on the expected impact of the new law:

> [I]t is made for cooperatives that are already big, to become bigger.
>
> (Cooperativas Agro-alimentarias rep.)

> Smaller cooperatives such as those you are telling me about [referring to MSCs], fantastic, but with that model, 4,000 cannot survive. Those that already have that model, fantastic, we provide them with certain services, this is not an issue that worries us. Those at risk are those that are not big and are not small. Right now they don't export but they are not able to sell all their production [. . .] we need to look at [those with] 500 members who are not ready to export and don't know what to do with their product, they are under-selling it. But the other ones, fantastic, we are not aiming to make all of them like Anecoop [. . .] apart that not all consumers are willing to pay for organic production, and here in Spain even less, so they are compatible.
>
> (Cooperativas Agro-alimentarias rep.)

> We have been trying for a long time that fewer ACs exist. For us it is not
> an achievement that there are 4,000 ACs . . . and if there were 5,000 it
> would be a disaster. What we are trying is to have 3,000, and if it is 2,000
> better and 1,000 even better.
>
> (Cooperativas Agro-alimentarias rep.)

The above comments suggest that efforts to reduce the number of ACs come
from both the government and the industry body that represents cooperatives.
The image of modernity for the sector is heading towards a future with fewer but
larger ACs that can compete in the European and international markets and have
enough power to negotiate conditions with large supermarkets. The following quote, however, reflects how embedded in the Spanish countryside the AC
culture is, and how far from this top–down modern vision the sector actually is:

> It is not a problem of the cooperatives, it is a problem of the agrarian
> structure of Spain, very, very atomised. But it is not a problem of the
> cooperatives. Who hasn't got a cooperativist granddad? Who hasn't got
> a granddad that owns half a hectare in a village? What is that good for?
> NOTHING! And then you have the Anecoop model that is a monster. So
> then I have these and these other ones. And then I have these in the mid-
> dle; my policies go to the ones in the middle. To the guy who's got a cow
> in the garage, against him, I can't do anything. To the big ones, in reality
> the policies should not bother them, they are going to serve them as small
> levers. But to those in the middle, I'm going to try, I shouldn't say it but
> this is what it is, or get them out of the market so they stop bothering, or
> get them to integrate in stronger market strategies.
>
> (Policymaker)

When asked to specify what they meant by the middle ACs "bothering", this
was the answer received:

> They distort because they don't take meditated business decisions in line
> with the market.
>
> (Policymaker)

> I'm not worried about Coren, Anecoop doesn't worry me, and if not,
> Sufrisa will buy it out or, they have directors much cleverer than all of us
> put together. The problem is the ones in the middle, like the farmers that
> are in the middle, that are slapped by the small ones because they are very
> flexible and by the big ones because they have "mucho cash". These small
> ones of the armchair (because they don't want to leave the armchair) are
> not going to work regardless of how much . . . the thing is, what money
> can I give them? It's very difficult to get to that level of policy, because
> Spain is stratified politically in three layers: the ministries, the autono-
> mous communities and the municipalities. I get to trick the autonomous

communities so that they coordinate their policies and we get 17 uniform policies. I don't reach these ones. People that make less than a million euros a year, they don't interest me, I can't reach them. We are in another game. This is very complicated. From here, we in the Ministry are seen as, well, and the ones in Brussels, we are at another level, they don't liaise with us. To start with, my policies are only for supra-autonomic coopera-tives,[10] because of the devolved powers.

(Policymaker)

The €1 million figure is not an official benchmark, but reflects the framing of what the Ministry considers to be its policy problem: medium-sized ACs. This battle to reduce the number of middle-sized cooperatives might, however, be ignoring the growing evidence that shows the negative effects of consolidation in the AC sector.

As discussed in the literature review, several authors have researched how the focus on getting bigger is affecting the social capital of members (Nilsson et al., 2012), who often perceive themselves more as customers than owners of their cooperative. But the loss might not be only social. Researchers studying the link between size and financial performance of agricultural cooperatives in Spain found no clear relation between these two variables. In fact, smaller farmer cooperatives show a better overall performance measured by several indicators: commercial activity, sales volume and better use of resources, both internal and external to the cooperative, such as public funding (Encina Duval et al., 2011). Meliá-Martí and Martínez also found in another study that the strategy of merging agricultural cooperatives produced no statistically signifi-cant improvements in the economic-financial indicators studied (Meliá-Martí and Martínez, 2014). These findings seem to be in line with early theories of cooperative degeneration, such as Hertzler's "cooperative dilemma", that sug-gest that the more successful cooperatives are economically, the more likely they are to fail socially (Hertzler, 1931).

Embedded politics in the rural world and ACs

Several participants spoke in different ways of the historical politics still embed-ded in Spanish ACs:

I can't be like Denmark and Holland. Holland is as big as Extremadura. But man, something in between, get together! What can't happen is what happens in Spain, that in every village there is a cooperative of the virgin and one of the saint, that some are red and the others are *fachas*,[11] that they don't get on well at all, probably commercialise the same prod-uct, pffff . . . they're both of first grade, they don't talk to each other but they're members of a second-degree one, that the second-degree one is good for nothing, the second-degree ones are good-for-noth-ing!.

(Policymaker)

> There are two historical moments in the creation of the cooperatives . . .
> that is because of Franco, those called after a virgin or a saint or from the
> first or second republic, called things like Cooperative of the Freedom, or
> Progress, and the ones in the *middleeee* . . . !
>
> (Policymaker)

> Yes, short-term policies are made, but this is the price we have to pay for
> having a democracy. In the North of Europe they have a different way of
> looking at things. It's the nature of Calvinism and Protestantism. In the
> South of Europe . . . everything gets politicised, and in agriculture even
> more, because agriculture is municipal politics. [. . .] In a village there are
> two bad things: the church and the bar.
>
> (Policymaker)

The allusion to the church and the bar in villages also raised the issue of how
politics and farming, and, by extension, ACs, are part of the wider public
debate beyond agriculture, farmers and policymakers. This topic of how ACs
are perceived externally was an interesting subject that emerged from the
research and it is briefly discussed in the next section.

Positive perception of cooperatives by the general public

Although it is not the purpose of this book to analyse the public perception
of ACs, it is interesting to briefly discuss this theme. At the end of 2013,
soon after the new law for the integration of ACs was approved in Spain,
Cooperativas Agro-alimentarias published a SWOT analysis of the sector in
order to inform the new National Programme for Rural Development agreed
at the Conference on Agriculture and Rural Development that took place that
same summer of 2013 (Cooperativas Agro-alimentarias, 2013). Among other
objectives, it was agreed this new programme would include measures to foster
integration in the AC sector. As part of this study, 2,000 interviews were carried out with consumers. Of the respondents, 30.6 per cent said they would be
willing to travel further to purchase food produced by ACs; 80.5 per cent also
thought that ACs could be a great channel to promote the Mediterranean diet
(Cooperativas Agro-alimentarias, 2013). These data suggest ACs are seen in a
positive light by the average consumer. However, the policymaker pointed out
that what consumers say is often different from what they actually do, either for
price reasons or for lack of information:

> [It] is a really good study, but we go back to the same thing, it's a study that
> says what consumers say what they would do, about the perception that
> the consumer has of cooperatives. [. . .] They like them. To the average
> Spanish consumer, cooperatives sound like something good.
>
> (Policymaker)

Cooperativas Agro-alimentarias tried to get a cooperative brand out a couple of years ago, but it didn't take off. [. . .] How many people know Anecoop are a cooperative? We assume the consumer has knowledge that they might actually not have.

(Policymaker)

Policy differences and similarities between Spain and the UK

In terms of policy contexts, four main similarities between the UK and Spanish policy arenas with regard to ACs have emerged from the document and interview data collected. Figure 7.3 summarises these parallelisms that are having an impact on the AC sector. At the international level, two key influences emerge from the data: multilevel governance and policy transfer.

1   Influence of MLG

One topic that most informants discussed at length was the Europeanisation process that shaped the development of the AC sector in both countries. This concept refers to the penetration of the EU into the political spheres of domestic policymakers, either through subsidy policies or directives or by changing their established patterns of decision-making and actor behaviour (Kohler-Koch, 1999; Clark and Jones, 2011). In the second half of the twentieth century, MLG brought a re-emergence of agricultural cooperation in the UK (late 1960s) and Spain (mid 1980s) as a driver to resist competition from and take advantage of the newly formed Common Market, strengthening export potential. Marketing cooperatives in the UK were not really needed until the 1960s, as marketing boards had statutory powers, and farmers were more worried

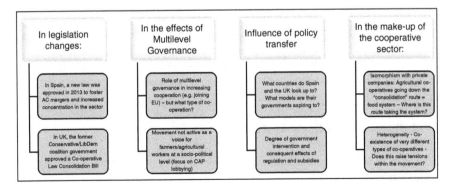

Figure 7.3 Spain/UK similarities in the policy contexts affecting the agricultural cooperative sector.

Source: Author

about production than marketing. This situation changed when the UK joined the Common Market, and, in 1967, the Central Council for Horticultural and Agricultural Co-operation (CCHAC) was created to assist farmers in marketing their products. CCHAC became Food from Britain. Support for agricultural cooperation in the UK was then halted in the mid 1980s. In the same decade, the Spanish Association for Agrarian Cooperatives was set up (only 3 months after the country joined the Common Market in 1986) as part of the strategy to enter the EU market.

Both British and Spanish public policies for ACs have been greatly influenced by the MLG effects associated with their EU membership. On the one hand, EU policies have promoted and subsidised the formation of member-owned enterprises in member states; however, the cooperative model is currently being affected by a conception of cooperatives that the EU is indirectly imposing through CAP subsidies. The EU is promoting a model for rural associationism that can gain acceptance in all member states (including new Eastern European countries that still have a distrust of the cooperative label owing to their own recent history). But that model is also conveniently unregulated and increasingly liberalised – POs. A PO can either be legally incorporated as a cooperative or not; it is not a requirement to access subsidies. This flexibility creates tensions within Cogeca, as discussed in Chapter 6, as long-established ACs feel threatened by their competition. As illustrated in the cooperative triangle (see Figure 10.1 in Chapter 10), it is important when studying any given producers' association to map and consider its legal and governance model as a way to understand its underlying rationale for cooperation and the objectives of the organisation.

At the same time, MLG has also continued to shape differences in agricultural cooperation in UK and Spain. Higher yields in Spain increase farmers' reasons to cooperate in order to access international markets, as the Spanish market cannot fully absorb domestic production. This is not the case in the UK, where there is a trade gap, as the country consumes significantly more fruit and vegetables than it produces. Internationalisation efforts take place in the UK with the aim of bringing food into the domestic market, rather than exporting it (Bijman et al., 2012), in absolute terms, although the UK government has been pushing for the internationalisation of British farming by increasing global trade through food exports (of mainly unhealthy food and drinks, such as whisky and biscuits (DEFRA, 2016).

2   Influence of policy transfer

The effects of policy transfer in the development of European agricultural cooperation are bifold:

- Transatlantic influence: One of the areas in which policy transfer has been noticeable is at the inter-sector level with regard to the solutions that have been attempted to tackle the problem of undercapitalised ACs: for example, transferring and adopting strategies from the private sector into the AC sector, such as the NGCs in the US that allow capital investments from non-members. NGCs (discussed in Chapter 3) were proposed in the late 1990s by a significant number of authors as a solution to the lack of competitiveness of the EU agricultural sector at the time.

- Cross-national EU influence: A second area of policy transfer is with regard to the UK government's and, mainly, the Spanish government's aspiration for their AC sectors to become as concentrated as the Dutch and Danish ones, and they have legislated accordingly to follow their example (CNC, 2012; MAGRAMA, 2012; EFFP, 2014).

The above influences have contributed to a change in the power dynamics of and impacting on ACs, as explained in Chapter 10 in a discussion of the double cooperative hourglass (Figure 10.3). These models of "competitive cooperation" have reversed the direction of pressure ACs used to exert outwards to protect members' interests to inwards, with ACs pushing their own farmers to accept the pressures and practices of agri-input companies, processors and retailers.

At the national level, two more influences have created increasing parallelisms in the development of the AC sectors in Spain and UK: legislation changes and cultural and political connotations associated with cooperatives.

3   Legislation changes

Whether affecting cooperatives across sectors such as in the UK, or specific legislation regarding ACs, the cooperative model seems to be back on the policy agenda of both countries, some say fuelled by the financial crisis and the search for alternative economic models. Chapters 2 and 6 have dealt in detail with these changes.

4   Cooperative is still a politically charged term

As discussed in the introductory chapter, ACs across the world have had complicated dealings with governments of different ideologies that, over the decades, have tried to promote them, control and/or shape them to achieve their own agenda. Despite the model's principles to be independent of government and creed, both left- and right-wing parties have realised the benefits of workers' cooperatives for decades (Coates and Benn, 1976). Factors such as land availability, ownership and cultural dynamics such as religion (Morales Gutierrez and et al., 2005; Spear, 2010) shaped the development of ACs in Europe. Within the continent, ACs still mean different things to different countries: in the UK, the term "farmer-owned-business" is commonly preferred and fuelled by the vast array of legal forms that cooperatives can opt for under UK law; in fact, the word "cooperative" does not appear in many ACs' names. Wilson and McLean found that many ACs try "not to look like co-ops as the co-operative identity is seen as 'old-fashioned'" (Wilson and McLean, 2012:537). In Spain, the government is offering groups of farmers not registered as ACs increasing support, and a common term used in policy documents is "associative entities", a label more encompassing than ACs (CNC, 2012; MAGRAMA, 2013). This reluctance to use the cooperative label can be explained by former failures of big cooperatives (especially in the case of the UK) and not wanting to be linked to the stereotype of cooperatives being unable to project a sufficiently businesslike image, increasingly expected from farming. In the UK, this reluctance also sprang from the wider context of a neo-liberal economy and lack of government support for ACs. Existing policy frameworks have sent contradictory messages for years, encouraging farmers to both compete and collaborate, as well as trying to transform farming at the same time into an increasingly "professional business".

As a result of all the above influences, an increased strategic isomorphism between dominant ACs and private agro-industries is palpable. Instead of breaking down power concentration in retailing and processing, policy-makers are keen to concentrate the farming sector through subsidies that encourage liberalisation of cooperatives and facilitate mergers. In summary, the above similarities in the policy contexts they operate in mean ACs are moving from operational approaches more focused on benefits for members and the cooperative movement into more strategic approaches. This move transfers more emphasis on to the enterprise dimension of their identity (competing both horizontally with other food companies and vertically with retailers), rather than on their social dimensions.

Nevertheless, it is fair to recognise that there is still huge heterogeneity in the sector. This heterogeneity not only refers to those thousands of ACs that are between those ranking in the charts and those alternative ones that have found their niches (about 90 per cent of ACs in Spain are in this in-between category). Although there is variety out there, the underlying productivist model is the same for most. However, there is a thought-provoking side to this heterogeneity. In Spain, for example, approximately 425 common-land cooperatives exist, where members co-own the land and cultivate it jointly (OSCAE, 2014). Several examples of cooperatives that link producers with consumers have also been found in both countries and are discussed in detail in Chapters 7 and 8. These multi-stakeholder models do not exist thanks to the food policy context, but despite it; these organisations are emerging with increased awareness of the risks of co-optation and regained efforts to protect the alterity of their cooperative models (through a variety of strategies presented in Chapter 8), sharing knowledge and joining efforts with other global social movements.

Conclusion

In order to delve deeper into the diversity of European ACs and the north–south divide, two ACs, one from the UK, Farmway, and one from Spain,

Anecoop, were selected as case studies. These cases reflect intense processes of consolidation in the AC sector through mergers and acquisitions. These farmer-owned companies are an example of the type of AC ranked in the top six in these countries in terms of turnover and size, and, as such, they are representative of the successful criteria of agricultural cooperation valued by the AC sector, national governments and at EU level. The chapter presented interview and document data categorised under the main themes that emerged from two types of actor: cooperative representatives and external experts (informants from government, industry AC representative bodies and civil society organisations). The AC sectors in Spain and UK share several similarities due to EU influence, including multi-level governance and policy transfer; cooperatives still being a politically-charged term and legal form; and recent policy changes at the national level.

Notes

1 A suitably qualified person is an animal medicines advisor, a legal category of professionally qualified persons who are entitled to prescribe and/or supply certain veterinary medicines under the Veterinary Medicines Regulations.
2 The Curry Report was commissioned by the UK government to analyse the factors that contributed to the foot and mouth crisis and the lessons learnt and make recommendations. This report is discussed in detail in Chapters 2 and 6.
3 A second-degree AC is a federated cooperative made up of first-degree producer cooperatives.
4 Utilised agricultural area includes arable land, permanent grassland and permanent crops.
5 POs are rural businesses, owned and controlled by producers and engaged in collective marketing activities. EU legislation explicitly states that a PO can adopt any legal entity. For a detailed introduction to POs, see Chapter 2.
6 This is the recent law approved to encourage consolidation in the Spanish AC sector.
7 The interviewee is referring to turnover of Danish cooperatives. This comparison with other countries is made in the preamble of the aforementioned Spanish law.
8 Mercadona is the largest retailer in Spain, accounting for about 22 per cent of food purchases.
9 The participant is referring to part-time farmers with minimal plots or herds, sometimes "keeping one or two cows in home garages".
10 Supra-autonomic cooperatives are those that operate across two or more autonomous communities. There are 17 autonomous communities in Spain, similar to county councils in the UK but with more devolved powers.
11 Many ACs in Spain are called after Catholic virgins or saints. *Rojos* (red) is slang for communists, and *fachas* for fascists, referring to the historical legacy of the Spanish Civil War.

8 Emerging models of cooperation in food and farming

Multi-stakeholder cooperatives

This chapter of the book is based on an article published in 2017 as part of the *Journal for Rural Studies'* special issue, titled "The more-than-economic dimensions of cooperation in food production", that had a special focus on conventional agricultural cooperatives and the complexity of processes and values included in cooperation (Emery et al., 2017). Two further case studies have been added to the discussion.

Multi-stakeholder cooperatives (MSCs) are a relatively new form of cooperative that has been emerging over the last two decades in Europe and North America (Lund, 2012). These cooperatives allow and bring together different types of membership – often consumers and providers of services and goods, but sometimes also workers and buyers (Kindling Trust, 2012). In Europe and Canada, MSCs are growing strong in social services and the healthcare sector (Münkner, 2004). In the US, the movement for relocalisation of food production and consumption has found a useful organisational and legal tool in the MSC model (Lund, 2012).

However, little empirical research has been done to explore and discuss how the MSC movement is developing new models of food production and provision. Furthermore, very little academic literature has dealt with MSCs in the specific context of food and farming initiatives, and existing publications focus on the US context only (Lund, 2012; Gray, 2014b).

Much has been written about alternative food networks (AFNs) and the potential of cooperatives to realise alternative food systems and reconnect consumers with producers (Kneafsey et al., 2008; Jones et al., 2010; Little et al., 2010; Goodman et al., 2011; Sharzer, 2012; Lawrence and Dixon, 2015). However, when it comes to cooperatives, most of the AFN literature focuses on consumer cooperatives, informal (not legally incorporated) buying groups and community-supported agriculture initiatives, rather than on farmers' cooperatives (Ronco, 1974; Allen, 2004; Little et al., 2010; Brunori et al., 2012; Smith et al., 2012; Cox et al., 2013; Caraher et al., 2014). ACs are large employers in the agricultural sector and enable farmers to reduce costs and gain lobby and negotiating power. Nevertheless, given the focus on competition and vertical growth in the AC sector discussed in Chapters 5 and 7, it is perhaps not strange then that the AFN literature does not dedicate much attention to the research and theorisation of ACs. Instead, the review of the literature presented in Chapter 3 revealed that the economics discipline has dominated the study of formalised cooperatives in agriculture (Helmberger and Hoos, 1962; Murray, 1983; Oustapassidis, 1988; Karantininis and Nilsson, 2007).

In this sense, despite the cooperative model often being recommended both in farming and also downstream at consumer level, a divisive line is still drawn between producer and consumer cooperatives. MSCs are emerging in alternative food systems as bridges trying to cover this gap. As opposed to conventional ACs made up of farmer membership only, the multi-stakeholder model brings together producers, consumers and/or restaurateurs in one single enterprise. This chapter analyses the emergence of MSCs, their dynamics and activities and the barriers they are facing to realise their own vision of sustainability, in social, environmental and financial terms.

As this model of cooperative is so rare and new, this chapter starts with an introduction to MSCs, providing an overview of the history behind them. In order to understand MSCs in the historical context of the cooperative movement, a review of relevant archive records, academic literature and current thinking on multi-stakeholderism in food and farming initiatives is provided. After discussing how the MSC model is not a new idea but is re-emerging in the context of globalised food and globalised social movements, data from four case studies are presented, two from the UK and two from Spain, to explore and discuss the dynamics and challenges facing these new cooperative arrangements.

The fourfold proposal for Open Cooperatives introduced in Chapter 4 is discussed and applied to the analysis of the case studies. Can these MSCs be considered Open Cooperatives? How do they reconcile the different interests of different groups within an organisation? Are they successful in serving the interests of the two weakest links in the food chain, that is, producers and consumers? Putting them in the context of the globally connected pro–commons movement, the Open Cooperative framework helps investigate whether the MSCs studied have the potential to connect with other pro-commons initiatives across the world, in an attempt to change, rather than adapt to, the food economies in which they struggle to survive (Gray et al., 2001). The chapter ends with a discussion of the dynamics and challenges facing these new cooperative arrangements, as well as the more-than-economic benefits they are reproducing through their practices by pushing the cooperative movement beyond survival mode in current market economies.

Multi-stakeholderism, an old idea coming of age?

It is important to acknowledge that multi-stakeholderism is not a new idea, and that early cooperators soon realised that bringing members together to cooperativise as many areas of their lives as possible made sense, at least in theory (Reymond, 1964). Historical data on the early attempts to create MSCs reveal the common underlying understanding shared by present-day MSCs of how cooperatives do not operate outside the market, but within it, and, as such, are strongly compelled to imitate capitalist relations as a way to survive in the dominant economic context within which they exist.

Before discussing MSCs in the current context, it is interesting to first review early cooperative attempts at multi-stakeholderism in the food system. As described in Chapter 2, the cooperative social movement began in the early nineteenth century with the realisation that the power of organised cooperation could have the potential to transform society and reverse structural conditions that produced great inequalities (Shaffer, 1999). Food has always been a core element in cooperativism since the very beginnings of the movement in the nineteenth century (Burnett, 1985; Birchall, 1994; Garrido Herrero, 2003; Rhodes, 2012).

Chapter 2 provided a detailed timeline of the origins of cooperatives in agriculture. In terms of food producers' cooperation, it was the Jumbo Cooperative Society, operating from 1851 to 1861 near Rochdale, that founded the first worker cooperative farm (Birchall, 1994). Both experiments, the Rochdale store and the Jumbo Farm, highlighted the active role of urban citizens in developing a new identity as workers, consumers and food producers. In this sense, Jumbo Farm could arguably be considered the first formal organisation of "cooperative prosumers", as engaged consumers also took an active role in food production (Pirkey, 2015).

Historical records from the National Cooperative Archive in Manchester (UK), which holds the collections of the Cooperative Union and the Cooperative College archives, provide evidence of how early cooperators soon realised the potential benefits and limitations associated with the possibility of merging different types of member into multi-stakeholder (MS) ventures. The topic of multi-stakeholderism was considered important and relevant enough to be discussed at several cooperative congresses (Reymond, 1964). The integration of different types of member into one single association is the main difference between MSCs and the more common single-membership cooperatives (Münkner, 2004). Conventional cooperatives are run by, and for the benefit of, their members, whereas the introduction of different types of member in MSCs both complicates and enriches the cooperative mission, as Lund has pointed out:

> MSCs are cooperatives that formally allow for governance by representatives of two or more stakeholder groups within the same organisation, including consumers, producers, workers, volunteers or general community supporters [. . .] The common mission that is the central organising principle of a multi-stakeholder cooperative is also often more broad than the kind of mission statement needed to capture the interests of only a single stakeholder group, and will generally reflect the interdependence of interests of the multiple partners.
>
> (Lund, 2011:1)

Efforts to create MSCs are not a new twenty-first-century invention. The UK's Cooperative Wholesale Society (CWS), founded in 1863 to supply the more than 1,000 consumer cooperatives already operating at the time in the UK, carried out one of the first attempts to bring together worker and consumer members; the challenges were soon evident, as this excerpt from an Economic and Social Consultative Assembly report reflects:

> For some time there were difficulties with the British CWS which had its own creameries in Ireland: was the purpose of the creameries to market the produce of the Irish peasant on the best possible terms, or was it to supply butter to British consumers at the lowest possible price? The conflict of interests led Plunkett and his colleagues to resign from the Cooperative Union and found the Irish Agricultural Organization Society in 1894.
>
> (ESCA, 1986:525)

The issue of "fair prices" is still largely unresolved today, as will be discussed when introducing the case studies in this chapter. Nearly a hundred years after the CWS was founded, one such attempt also took place in France in 1959, through an agreement between the national agricultural and consumers' cooperative organisations. A commission was set up to report on the difficulties encountered, summarised as follows (Reymond, 1964):

1 The system proved unwieldy for handling operations through the central organisations and it was recommended that the largest number of transactions were better carried out locally.
2 A process for ensuring compliance with quality standards and agreed prices had to be improved.
3 Price-fixing was a long-standing problem, and despite being in a closed, cooperative "full circle", ignoring the normal market to negotiate prices proved impossible.
4 Price variations complicated the purchasing side of the relationship.
5 There was a fear of damaging existing relations between suppliers and dealers.

The above points highlight how cooperatives do not exist in a policy or economic vacuum, but, as today, struggle to survive in capitalist societies ruled by the laws of the market. Nevertheless, the French commission also noted how the will to succeed on both sides was a significant strength of the model. The ideal endured, at least at the theoretical level, and more modern cooperative thinkers continued to write about the economic benefits that they identified would occur when production and consumption were linked:

> [I]f a considerable proportion of farm crops could be sold directly by farmer-owned enterprises to consumer-owned ones, the "spread" between what farmers receive and what consumers pay would amount simply to the costs of processing, transportation and sale.
>
> (Voorhis, 1961:83)

Despite the apparent conflicting interests from producers/providers and consumers that MSCs try to reconcile, frequent cases of cooperatives extending their membership, and thus the range of objectives in the membership, currently exist in other sectors than food. An example of this is the case of mutuals such as saving cooperatives that include members with completely opposite interests, i.e. savers and borrowers (ResPublica, 2012); but it has been argued that interest harmonisation between savers and borrowers is facilitated by the fact that, over the course of time, most members turn from depositors to borrowers, and vice versa (Münkner, 2004). This change of identities is not as common in food, although all farmers are also consumers, and, in MSCs – as the case studies presented below will illustrate – members are given opportunities to embody different identities as consumers, producers, workers or volunteers (Kindling Trust, 2012). Leviten-Reid and Fairbairn have also challenged the negative predictions that transaction cost theories make of MSCs and have proposed a new framework based on a "governance of the commons" theory, to show how the MS model can be efficient and effective (Leviten-Reid and Fairbairn, 2011). Theories of the commons normally refer to the challenges of managing common-pool natural resources, such as rivers, fisheries, forests and shared irrigation systems (Ostrom, 1990). However, this model has also been applied to worker cooperatives in capitalist societies, proposed by some authors as "labour commons" that generate commonwealth through their practices (Vieta, 2010).

In this context of the commons, evidence from Italy, the first country with MS cooperatives providing social services since 1991, shows that an MSC is not a zero-sum game – one cohort of members does not need to win to the detriment of others (Borzaga et al., 2010). In this context, Mooney has highlighted how the ongoing "rationalisation of an antagonistic economic relationship in its formulation of 'producer groups' and 'consumer groups' who simply carry on the battle in another sphere" is divisive (Mooney, 2004:86); it is also against the original cooperative vision that aimed to create an organisational structure that could cover interests and needs for a common good.

Münkner has called for the introduction of MSCs as more efficient and locally embedded providers of public services, but also because they represent "new and attractive forms of cooperation in times where the numbers of registered cooperatives are steadily shrinking as a result of mergers" (Münkner, 2004:50). For Münkner, the fact that MSCs are emerging all over the world is a sign of how conventional rules of cooperation are outdated and are being reinvented to "maintain organised self-help as a relevant answer to current problems in times of rapid change" (Münkner, 2004:65).

Along similar lines, Michael Bauwens, the founder of P2P, has also stated the limitations of current cooperative models:

> The problem with the capitalist market and enterprise is that it excludes negative externalities, [both] social and environmental, from its field of vision. Worker-owned or consumer-owned cooperatives that operate

in the competitive marketplace solve work democracy issues but not the issues of externalities. Following the competitive logic and the interests of their own members only, they eventually start behaving in very similar ways.

(Bauwens in Johnson, 2014)

Other cooperative thinkers share Bauwens's concerns over the social and environmental externalities ignored by private companies and how cooperatives are copying the same steps in order to be able to compete financially. In 2014, Thomas Gray called for the formation of MSCs as a way to overcome three key historical cooperative tensions: (1) participation and democracy versus efficiency and capitalism, (2) localism versus globalism, and (3) production versus consumption (Gray, 2014b). Gray suggests that MSCs can help ease these tensions while also offering an integrative organisational structure that can automatically internalise the current externalisation of the human and environmental costs involved in food production and consumption (Gray, 2014b). These tensions are similar to the ones earlier identified by Mooney (2004).

Lund has also studied modern MSCs in food and farming in the context of the emerging concept of value chains (Lund, 2012). In contrast to the dominant notion of the supply chain, the development of the concept of the value chain provides a framework for indicators beyond economic transactions; this is a key consideration in food production and consumption, as these activities involve many cultural and social aspects, such as taste, identity, connection with nature and community, and so on, that are ignored in financial exchanges (Baggini, 2014). As opposed to other actors who predict unavoidable conversions of MSCs to private firms owing to complex governance structures and cumbersome decision-making processes (Lindsay and Hems, 2004), Lund offers a rationale for seeing membership heterogeneity as a strength rather than a barrier to efficiency (Lund, 2012). By fostering long-term relationships rather than punctual commercial transactions, Lund affirms MSCs can be transformational and overcome the higher transactional costs that traditional economic theory would expect from the involvement of several parties (Lund, 2012).

Like Lund, Bauwens draws on the idea of "value chain" but goes beyond Lund and Gray by calling, not only for MSCs, but for a more specific new model that has been labelled Open Coops. This model combines multi-stakeholdership and the co-production of the value chain by everyone affected by a provisioning service. Bauwens believes MS is the cooperativism of the future and can help overcome co-optation trends in conventional cooperatives. The Open Cooperative framework is introduced in the next section, then investigated throughout the chapter and used to analyse the case studies presented.

The Open Cooperative framework

So far, the ACs discussed in Chapter 7 had a conventional membership with only one category of member: farmers. This current chapter attempts to move the analysis of ACs a step further by studying cases of MSCs. For this purpose, the need for a more specific theoretical lens was identified. The Open Cooperative framework introduced in Chapter 4 is considered in more detail in this chapter as it emerges from a pro-commons, globally connected social movement as a lens to analyse the social movement phenomenon of which the MSCs discussed are themselves a part (Bauwens, 2014; Bauwens and Kostakis, 2014).

The Open Cooperative framework is relevant to more-than-economic benefits of cooperation because it highlights the roots and links MSCs have with civil society organisations and other global social movements such as the open data and open economy communities, solidarity economy, food sovereignty and organic movements (REAS, 2011; Bauwens, 2014; Cooperativa Integral Catalana, 2016; Manchester Veg People, 2016).

The open economy movement fights the increasing privatisation and commodification of knowledge, especially in the context of the internet age (P2P Value, 2016). Open economy activists are working to develop commons-based models for the governance and reproduction of abundant intellectual, immaterial resources (e.g. software, apps, etc.). At the same time, the P2P Foundation is working to link up with the cooperative movement, as it sees cooperatives as the ideal organisational type to develop a reciprocity-based model for the "scarce" material resources we use for reproducing material life (Bauwens and Kostakis, 2014). The vision is for the surplus value to be kept inside the commons sphere itself, creating a merger between the open peer production of commons and cooperative ways of producing value: "it is the cooperatives that would, through their cooperative accumulation, fund the production of immaterial Commons, because they would pay and reward the peer producers associated with them" (Bauwens and Kostakis, 2014:358). In this context, the Open Cooperative framework calls for the evolution of the conventional cooperative model across four simultaneous dimensions (Bauwens and Kostakis, 2014):

1   Open Cooperatives should include their objectives in their own statutes, and their work should be aligned towards the common good, integrating externalities into their model.
2   All people affected by the activity should have a say (this is the specific multi-stakeholder nature of Open Cooperatives), practising economic democracy.

As Bauwens has pointed out, these two characteristics already exist in the solidarity cooperatives – which is another name used in the literature to refer to multi-stakeholder cooperatives (Lund, 2012) – such as the popular social care MSCs in Italy and Canada. The P2P Foundation framework advances two extra practices that MSCs have to incorporate in order to become meaningfully transformational (Bauwens and Kostakis, 2014):

3   The cooperative must co-produce commons for the common good, whether immaterial or material.

4   The final requirement is a global approach, to create counter-power for a global ethical economy consisting of cooperative alliances and a disposition to socialise its knowledge.

The next section introduces the four case studies from the UK and Spain; data are then discussed in the framework of Open Cooperatives, followed by an analysis of the theoretical and practical implications for ACs and farming policy.

Case studies

In this chapter, four case studies, two from the UK and two from Spain are presented. As discussed in Chapter 6, these two countries have a remarkable set of differences and similarities (historical, political, economic, etc.) that offer an interesting combination to study agrarian cooperativism from many perspectives. In Spain, collaborations (some informal, some more structured) between consumer associations and groups of farmers have been going on since the 1980s and have been discussed in academic literature (Alonso Mielgo and Casado, 2000). Multi-stakeholder experiments in Spain have grown exponentially since the beginning of the financial crisis in 2007–8 (Calle Collado and Gallar, 2010; Saravia Ramos, 2011). In the UK, the literature on MSCs is pretty much non-existent, but initiatives based on MSC values are also emerging (Cultivate Oxford, 2016), although only one food and farming cooperative legally registered as an MSC was found (Manchester Veg People, 2014) and was taken as a case study. The other UK case study is actually legally registered as an agricultural workers' cooperative, but, given its strong links with its consumer members and the rarity of its legal form (as there are hardly any worker cooperatives in the farming sector), it is considered an MSC for the purpose of this analysis. The next section introduces the four case studies from the UK and Spain, and data are then discussed in the framework of Open Cooperatives.

UK case studies: Manchester Veg People and OrganicLea

Case study I: Manchester Veg People

Manchester Veg People (MVP) is an MSC based in north-west England, the cradle region of modern cooperation and also home to Co-operatives UK, the representative body of cooperatives in Britain. Manchester is one of the UK's largest cities, with a population of 2.5 million people; despite its size, before MVP started trading, Greater Manchester had one of the lowest levels of access to locally produced food in the country, as well as being situated in the region (the north west) with the lowest number of organic farmers (Kindling Trust, 2012).

MVP was the result of an ongoing collaboration that started in 2007 between the Kindling Trust and a small group of producers and two buyers that were exploring how best to coordinate their demand. Their aim was to develop "a new model for the local food supply chain", MVP's strapline being: "Keeping it fresh, organic and local". At the time of writing, MVP's membership comprised five growers, 20 buyers (including restaurants, caterers and public sector organisations) and four worker members. Apart from one producer who is more than 60 years old and a third-generation farmer, all MVP's other growers are younger people who are new to farming, a remarkable fact considering that only 12 per cent of UK farmers are aged between 25 and 34 years old and around 41 per cent are more than 50 years old (LANTRA, 2011).

Although the project started with no funding, MVP then received financial support and advice for the development of the cooperative, especially around research to develop the structure, membership documents, policies and other background material. The cooperative received £23,500 from the UK Lottery Making Local Food Work programme.[1] This sum funded the development process, the coordinator's salary and other costs, including early distribution costs, branding development and website, packaging, registration costs and the writing of a report to document the set-up process and share best practice (Kindling Trust, 2012). The Rural Development Programme for England (partly financed by the European Agricultural Fund for Rural Development) awarded MVP a grant for a coordinator for 18 months and 40 per cent towards capital items to help expand the range of the products offered through the cooperative. Other supporters that worked with MVP and helped it through its setting-up stage include the Cooperative Enterprise Hub, Plunkett Foundation and Esmée Fairbairn Foundation (Manchester Veg People, 2014).

MVP self-describes on its website as "something different". It is indeed so, as MVP is the only MS food cooperative in the UK that links farmers and buyers. MVP was originally registered as a company limited by guarantee, owing to the constraints that the withdrawable share limit of £20,000 per member posed (this limit was national policy at the time but was raised to £100,000 in April 2014). MVP's early engagement with Co-operatives UK and Making Local Food Work (a 5-year Big Lottery-funded programme run by a consortium of organisations to improve the sustainability of community food enterprises) expanded its network in different directions, and it was soon referred to Somerset Cooperative Services (SCS) to discuss its vision of forming a cooperative with weighted voting. In 2009, SCS, a community interest company based in the south of England, became a sponsoring body able to register industrial and provident society cooperatives with a new model of cooperative rules, also known as the "Somerset Rules". These rules provide cooperatives with more flexibility than other existing model rules as they enable a heterogeneous type of membership while strictly adhering to all the cooperative principles (Somerset Cooperative Services, 2009). In MVP, the voting is weighted as follows: 45 per cent for growers, 30 per cent for buyers and 25 per cent for workers. Weighted voting avoids replicating the power imbalances characteristic of the conventional food system; so far, there has not been any need to vote, as all decisions have been reached by consensus decision-making processes. Prices are calculated by adding a 35 per cent mark-up to cover the running costs of the cooperative to the cost of production of each crop (including seasonal variations).

As opposed to other local food schemes and cooperatives that buy produce from non-members to ensure they meet demand and have a more varied offer, when MVP started, it did not buy in from non-members. It works to support buyer members to create more seasonal menus, but buyers can and still do use other suppliers, which, as will be discussed in the next section, can create difficulties for MVP's growers owing to the cost of dealing with small orders. Recently, MVP has started to sell to non-members (at a higher cost) in order to increase demand.

In 2014, MVP and the Kindling Trust received funding from the charity Ashden Trust to carry out a pilot project with three public sector organisations over a year, with the aim of introducing more local organic vegetables on their menus within existing budgets (Manchester Veg People, 2014). Supplying public institutions is one of its main strategies to scale up and democratise access to both organic and local food. MVP also serves the University of Manchester – its biggest buyer so far – which became a member in 2010; the university started ordering for one of its kitchens, and, at the time of writing, MVP served salad leaves for all its halls of residence plus 18 of its 28 campus outlets. The unique combination of local and organic defines both its identity and also the main selling point of MVP in Manchester. By only selling locally grown produce (defined as within 50 miles of the city centre, meaning short travelling distances), plus the fact that the produce is picked to order the day before delivery, MVP guarantees freshness, an attribute that translates into quality but also into a longer window of time for chefs to use the produce, thus reducing waste.

In summer 2015, in addition to this stream of work with farmers and restaurateur buyers, MVP started a parallel venture selling veg boxes directly to consumers. The boxes are delivered to three hubs to save transport costs, and consumers collect them from there. This part of the enterprise does sometimes sell produce from non-members in order to provide a more comprehensive range of vegetables on offer. It aims to incorporate fruit and other products, such as dairy items, in the future.

Case study II: OrganicLea

OrganicLea is a workers' cooperative based in North London following organic and permaculture principles and fostering strong worker and stakeholder engagement as part of its objectives. OrganicLea's workers' cooperative status is a rare legal form in UK farming, one of the very few of its kind in the country. The combination of its governance model and production approaches makes OrganicLea a fairly unique case in UK agriculture.

The OrganicLea project started back in 2001 on an acre of once-derelict allotment land in the valley of the River Lea, which inspired the cooperative's name. This valley has a long farming tradition, as for centuries settlers have made the most of the fertile soil by the river, and the area became known as the bread basket of London. Being based in Greater London, OrganicLea's vision from the beginning was to produce more "food for the city, in the city". Originally, OrganicLea had mainly a social focus, hosting events and courses and operating as an allotment, selling surplus only occasionally.

But, in 2006, following a desire to contribute to and bring to reality the newly born concept of the "local food economy", the Hornbeam Centre and Café was opened, becoming one of the first local food hubs in the country, and regular trading commenced with its weekly market stall. The activities of the centre developed even further after funding from the Big Lottery's Making Local Food Work programme was received in 2008. New initiatives included the still-ongoing weekly veg box scheme (that, at the time of writing, had about 380 subscriptions) and the innovative Cropshare scheme that enabled local gardeners and allotment holders to sell their surplus to the project's outlets, including three market stalls. Roughly at the same time, OrganicLea took over the neighbouring plant nursery formerly owned by Waltham Forest Council, which vastly increased the range of crops that could be grown and traded. It estimates that about 160 volunteers and trainees learn horticultural skills on its farm every year. The cooperative employs 17 members, who work in production, distribution and training. Additionally, it also creates paid employment through its extensive outreach programme running food growing sessions at local schools, children's centres, community centres, sheltered accommodation sites and housing associations (Land Workers' Alliance, 2015).

Nowadays, OrganicLea continues to emphasise the fact that its work in agriculture goes beyond food growing, and it synthesises its wide range of activities in four main interrelated streams: "We grow food", "We sell food", "We help you grow" and "We work for change". Its broader policy and political work and activism involve connections with wider networks, such as the Community Food Growers Network and the Land Workers' Alliance, the new union that represents small- and medium-scale farmers who follow environmentally friendly growing practices. OrganicLea is also part of the international food sovereignty movement, a connection formalised through the Land Workers' Alliance's membership of Via Campesina.

In 2015, OrganicLea set up a "FarmStart" project to support new entries into farming, following the lead of the Manchester-based Kindling Trust. FarmStart Manchester was the UK's first such initiative, based on successful models in Canada and the US that aim to help people aspiring to make a living from organic farming, or what OrganicLea refers to as: "growing new growers" (OrganicLea, 2015). OrganicLea has received funding and support from the Esmée Fairbairn Foundation, Enfield Council and Haringey Council for this project, which aims to create at least six new incubator plots for new growers. As will be discussed later in the chapter, OrganicLea's case intertwines issues of access to land, feeding growing cities and the potential role of cooperativism in food and farming.

Spanish case studies: Actyva and Esnetik

Case study III: Actyva

Actyva is an MS initiative based in Cáceres, a city in Extremadura, a region located in western Spain. Extremadura is one of Spain's 17 autonomous communities, with a mainly rural character, the fifth largest by area but the twelfth in terms of population numbers, with only just over 1 million inhabitants.

The idea for the Actyva project originated in 2010–11 when members who were unemployed and at risk of exclusion from the National Confederation of Labour (CNT) – a long-established union based on anarcho-syndicalist principles – decided to create a formal structure that could help them meet, not only their own needs, but also those of people around them. Extremadura has the highest rate of undeclared or shadow economy in the country at 31.1 per cent (6.5 percentage points higher than the national average), with more than 30 per cent unemployment (GESTHA, 2014). Actyva aims to offer both a channel to help people out of the "submerged economy" and a model that can be replicated.

As opposed to MVP, Actyva did not receive any external funding for its development and relies on members' joining fees and investments; however, Actyva does deliver information about subsidised training courses to members. The cooperative aims to realise a model of local sustainable agriculture that follows agro-ecological and food sovereignty principles, based on mutual aid and networking and aims for social, economic and environmental sustainability. Spanish organic farmers have found bigger and more lucrative markets for their produce in northern Europe (Pratt and Luetchford, 2014:133), but Actyva's mission is to keep agro-ecological produce local, only exporting to Europe surplus products with saturated markets in Spain. Another condition for export is only selling to like-minded, politically aware buyers that are aligned with solidarity economy practices.

Actyva's two main food initiatives are Big Brother Bio-Farming (BBBF) and "*Cáceres para comérselo*" (which could be translated as "Caceres, so good you could eat it". BBBF, also called by some members "the big brother of organic farms", is an online platform that facilitates and encourages small, organic producers to live-stream activities happening on their farms, as well as providing online courses and spaces for consumers and producers to create and develop regular contact. The name not only refers to the fact that consumers can watch what happens on the farm, but also plays with the traditional Spanish street retail expression of the three Bs: "*bueno, bonito y barato*", which translates as "good, nice and cheap", challenging the assumption that organic food has to be expensive or only for "discerning customers". In conjunction with BBBF, "*Cáceres para comérselo*" was launched in March 2014 with the aim to support small, local agro-ecological producers in their marketing and distribution activities. The aim of this programme was to sign up 300 households to receive a weekly vegetable box and to become members. Deliveries were going to go, not only to households, but also to members' workplaces, in order to concentrate demand, increase the impact of sustainable consumption and reduce delivery costs and emissions. The initiative was paused at the end of 2015, as the target had not been reached; sales resumed in May 2017 with a new online shop, a physical venue for collections and a wider offer of refrigerated products.

Case study IV: Esnetik

Esnetik is a not-for-profit MSC based in the Basque Country. Esnetik was formed as a response to the marginalisation of local traditional shepherds who were being locked out of their conventional routes to sell their milk (Ajates Gonzalez, 2017d). These shepherds were being dropped by larger milk-collecting companies or cooperatives, either because they were not on a main collection route or because they focused too much on milk quality rather than quantity. Esnetik started selling sheep-milk products in May 2012. It continued to grow and, at the time of writing, the MSC employs three full-time workers and a part-time driver for deliveries and collections. The meaning of the name Esnetik comes from the Euskera language and is made up of two words: *esneki* (dairy) and *etika* (ethics). Esnetik likes to present itself as a cheese composed of the following slices: its diverse membership, philosophy and traditional food preparation methods. The members of the cooperative are shepherds, consumers, workers and collaborating organisations (a combination of NGOs, local authorities and rural development organisations that were approached with the aim of bringing closer together the urban and rural dimensions of food production and consumption), around 200 people in total. At the time of writing, there were five shepherd members of Esnetik. This MSC buys 100 per cent of its production at a fixed, fair price agreed with the shepherds, and they all receive the same price, regardless of the volume produced or their location; this was in contrast to their previous situation, where they were offered very low prices if they were off the main collection route and, in some cases, could not even have their milk collected. Some of the producers milk by hand and, in general, have a traditional method of production that does not fit the industrial model that values the number of litres as it lowers transport and processing costs.

Esnetik sells as much of its produce (cheese and yoghurt) as possible to its consumer members (individuals and consumer groups), and the rest gets sold to a milk parlour; Esnetik covers the price difference between the price agreed with the producers and the price the milk parlour is willing to pay. Its objective is to grow its network of consumer members so that demand is enough to process more milk in Esnetik and sell less to the milk parlour.

Five members form the management board: two people representing producer members, two representing consumer members, plus a social movement group representative. At the time of writing, the supporting organisation with a representative sitting on the board was the Basque farmers' union EHNE Bizkaia; this union was instrumental in making the Esnetik project into a reality. Additionally, other civil society organisations and local authorities are also supporters and members of the cooperative.

Esnetik has close links with two movements that also inform its practice and operations: the food sovereignty movement and REAS, which is the *Red de Economía Solidaria y Alternativa* (Network for Alternative and Solidarity Economy). Being a cooperative, it was easy to assume that the cooperative principles would be core to Esnetik; however, in order to avoid wrong assumptions, participants were asked to express what movements and principles they identified with more. It was interesting to learn that REAS has six principles that, along with feminist and food sovereignty ones, are more central and core to the *raison d'être* of Esnetik than the cooperative principles themselves.

The cooperative fosters diversification instead of specialisation, an approach that supports new people with fewer resources moving into agriculture: Esnetik has noticed that a key barrier to entering the livestock sector is the large amount of money that new entries have to invest. By not pushing shepherds to increase quantity but fostering instead diversification and production of diverse crops for self-consumption, Esnetik promotes an agro-ecological model that can help new producers make a living in a more sustainable way.

Can these MSCs be considered "Open Coops"?

This section analyses the findings from the four case studies presented. The following analysis is structured according to the P2P Foundation's four requirements for "Open Coops" discussed at the beginning of this chapter, namely: integrating externalities into their model, practising economic democracy, co-producing commons and having a global outlook (Ajates Gonzalez, 2017a). Each heading discusses a selection of the most relevant practices from the case studies rather than a list of practices for each of the MSCs presented. Later in the chapter, the associated theoretical and practical implications for ACs and farming policy will be examined.

Internalising negative externalities

Financial externalities

Chapter 1 discussed how farmers get an increasingly small percentage of the profits generated in the food supply system (DEFRA, 2012). Pricing is a thorny issue, not just for MSCs but for the whole sustainable food movement (Pratt and Luetchford, 2014). The debate about the idea of a "just price" has been going on since at least the time of Aristotle (Pratt and Luetchford, 2014:34). The multidimensional character of sustainability and the multiple meanings of food mean that fixing work, inputs, impacts (environmental and socio-economic) and values into a price is a daunting task with no easy method (Sustainable Food Trust, 2013). MVP's attempt to develop a still outstanding pricing formula has proved to be a task more challenging than expected. Nevertheless, by having this debate, these MSCs are redefining and relocating value; raising questions about who enjoys the benefits of its creation; questioning whether that value is economic, social or other (Pratt and Luetchford, 2014:13); and disentangling the significance of money and monetised value in closer trading relations.

Despite being paid a fair price for their products, beginnings are hard, especially for farming businesses, and, as one of the MVP members points out below, despite efforts to coordinate joint deliveries with other members, there was a tension between their vision and the costs of some of the initial deliveries:

> The orders from MVP have been very small, so when we've been driving them to the unit, which is the other side of Manchester is actually costing us because there is nearly nothing at the back of van, but we do it because we are committed to the vision. [. . .] Once we get to the kind of volumes that there's potential for, then suddenly we'll have £1,000k at the back of the van rather than £60.
>
> (MVP member)

Actyva in turn is trying to go beyond conventional capital by being involved in the alternative currency movement and being a member of the Community Exchange System used by many Spanish MSCs (Red de Cooperativas Integrales, 2016).

Another difference in the case studies refers to their approach to organic certification. MVP's produce is organic certified, plus their buyers also display MVP stickers in their premises to communicate their participation in the cooperative to consumers. This raises both their returns and prices and also gives them a competitive edge as the only producers of local organic food in Manchester. In contrast, Actyva opted for a Participatory Guarantee System. Each strategy comes with associated benefits and risks. As discussed earlier, information about the invisible attributes of food is increasingly important for consumers, and, as a result, this information has also become a selling point that needs to be communicated somehow. From a sociological perspective, Becker has discussed how physical objects get their character and meaning from the collective activities of people (Becker, 1998), conveyed, in the particular case of MVP, through stickers. A humble leek suddenly changes its value and connotations when it carries a sticker that says "organic" or "MVP", as it conveys political, geographical, socio-economic and quality dimensions that allow buyers to express their values when buying food (Kneafsey et al., 2013). On the other hand, the sticker becomes a replacement for trust developed through multiple encounters allowed by short distances and familiarity, allowing "conversion of culturally defined values into monetary value" (Pratt and Luetchford, 2014:40; Ajates Gonzalez, 2017c).

At the end of 2015 Actyva was reflecting on its *Cáceres para comérselo* project, as it did not reach its target number of orders to make it financially viable. It considered several options, including: supplying restaurants and a catering company serving school meals; resuming the negotiation of a space in the city centre's wholesale market with its local authority, with the idea of using it as both a storage and distribution point; establish a monthly market; and, finally, trying to match the offer to consumer members' needs more closely. Their sales resumed in May 2017, opting for both physical pick up points and introducing more chilled products on offer.

Environmental externalities

The ILO has published emerging evidence that suggests that, where cooperatives are multi-stakeholder, the capacity for businesses to push negative environmental externalities (i.e. waste and pollution) upon particular stakeholders is diminished (ILO Co-operative Branch, 2012). The topic of fair prices just discussed opens up the complex topic of sustainability, as the conventional agriculture these MSCs have to compete with treats human and environmental costs as externalities and thus remains more competitive in terms of price (Gray, 2014b). As suggested by Gray, MSCs offer an organisational structure that automatically internalises human and environmental costs (Gray, 2014b); however, by internalising these costs, they create other tensions related to their financial sustainability, as we can see from those MSCs that are struggling to grow their enterprises and remain financially viable.

When analysing these MSCs' approach to sustainability, a dual strategy emerges: the MSCs try, first, to minimise and, second, to internalise normally non-accounted externalities of food production through their growing methods, governance and close connections with their buyers and locality to reduce waste and transport. Their focus on local and community knowledge and ecological sustainability based on long-term views, rather than short-term profits, risks their financial sustainability, as they have to compete with other players who do not cover those externalities (Böhm et al., 2010).

Aspirations of autonomy are intrinsic to Esnetik's understanding and practices of sustainability. Sustainability and autonomy are linked in the way Esnetik members understand organic agriculture, but also in their direct relation with groups of consumers. For Esnetik, organic production is a synonym of autonomy. Esnetik farmers oppose organic production methods that are based heavily on reliance on external inputs, as this type of organic agriculture is considered a trap that does not change producers' dependency on agri-input industries. At the other end of the supply chain, Esnetik also refuses the type of organic production relying on large supermarkets for routes to markets, developing instead a network of trusted buyers (either consumer groups or small, like-minded retailers); this approach offers them independence from large distributors, as this quote reflects:

The biggest learning that organic livestock has given me is the capacity for autonomy it granted me. If we don't understand that organic farming, that agroecology, are means, tools for the autonomy of farmers, to produce at lower cost of production, then I think we are getting it wrong. [. . .] That is the problem . . . that a new organic agriculture is being made [. . .] just as dependent as the other one.

(Esnetik member)

When one of Esnetik's members was asked what sustainability meant to them, they answered:

There is a lot of debate, to me [sustainability] is what allows the producer in this moment in time to live with the maximum degree of autonomy on the one hand, and to perpetuate in time the continuity of the farm. I prefer not to enter into details, for example, around local produce, zero-km products, sterile debates from my point of view, that at the end of the day, large retailers take advantage of, because they are able to absorb them quickly and in fact they are already absorbing them and local products are part of large retailers marketing. And for that reason I say, myself who am in that struggle, we need to create a complete distribution that becomes an alternative way of consuming for people who want to participate in this process.

(Esnetik member)

The above two quotes reflect producers' awareness of the risk of co-optation associated with narrow definitions of sustainability based on simplistic metrics. At the time of writing, some of the members of Esnetik were starting a new, separate cooperative that includes more products beyond dairy ones (e.g. oranges and olive oil from the south of Spain) from like-minded producers. The rationale for this separate cooperative was to be able to provide consumer members with a serious alternative to supermarkets by offering a wider range of products and thus covering more needs. These conceptual tensions were bringing up differences among members who had more purist views of sustainability based on localism, which are complicated in a country such as Spain, where the climate, the regions and the crops are so varied. This final point reflects how definitions are constantly evolving, benchmarks are changing, and consensus is hard to maintain.

For MVP, reducing waste both on the farm and downstream in the food chain is one of its main objectives, something it achieves by investing time and resources in maintaining regular discussions between growers and buyers around crop planning, picking to order to ensure the produce is as fresh as possible and working on adapting menus to seasons. In turn, Actyva attributes similar importance to social and environmental dimensions of sustainability, reflected in its BBBF initiative and its determination to work with members at risk of social exclusion. OrganicLea reduces environmental externalities by adopting permaculture principles and growing methods.

These cooperatives are putting efforts into sustaining and extending the farming population by working with growers new to farming. Kindling Trust's FarmStart project is supporting new entries into agriculture. The FarmStart initiative was implemented in London through connections with OrganicLea (OrganicLea, 2015).

Reconnecting: economic democracy

Food is rich in social meanings, not just nutrients (Ajates Gonzalez, 2017c), and the importance of making consumers aware of the invisible attributes of food (method of production, distribution, trading conditions and how beings and the planet were treated along the way) is patent in the case studies. The industrial food system is often described as "disconnected"; however, as Kneafsey and colleagues have pointed out, never before have retailers been so connected to producers and processors through contract farming arrangements and traceability systems (Kneafsey et al., 2008). The cooperatives discussed in this chapter have identified the areas of the food system that actually need reconnecting, reflected in their attempts to bring together the two weakest links in the food system: small producers and consumers.

As commercial organisations, MSCs recreate parallel versions of the conventional market, creating spaces where producer–consumer relations and expectations must be negotiated, agreed and managed. The case studies challenge the accepted message reinforced by supermarkets that assumes the aims of farmers and consumers are irreconcilable. Furthermore, they call for connected "local to local" networks of place-based food systems, rather than reinforcing the idea that there is a single integrated food system. MSCs aim to go beyond simplistic dichotomies of local and global scale and of urban and rural dimensions. However, trying to convince others that multi-stakeholderism is a good idea is not easy, as these quotes reflect:

> When we originally had the idea people said it's really hard to even get growers together, never mind getting growers and buyers work together and everybody I spoke to said "no, no, it doesn't make any sense, X needs to set up two different coops, a producers coop and a buyers coops" and I was saying "but that doesn't make any sense because then you've got total conflict of interest going on and the only way you can sort that out is to bring people together", and it's just, I don't know, it's funny. [laughs]
>
> (MVP member)

> People say you're mad if you try to reconcile the interests of producers and consumers because people say food producers want high prices and consumers want low prices. Our research has shown that their interests are really really tightly aligned. And we are told they're very different, hmm.
>
> (Plunkett Foundation representative)

When asked by whom, the following long but illustrative reflection was offered:

> [I]t tends to be the retailers and the processors, the people who keep them apart [. . .] people just assume that they have very different views and the cooperative movement is based on the fact, the belief that if you are in a rural community, producer cooperatives are your base because that is where your farmers are and consumer cooperatives operate in urban areas because that is where the consumers are and that is quite wrong really, that is one of the great regrets, and our founder talked about it, they did not ever reconcile that, they didn't ever get to a common understanding that you can meet producer and consumer interests through cooperative action. There are still today very few examples of that, which is a shame, is a shame, it makes sense. In theory it makes a lot of sense, but I think people would argue that in practice is pretty difficult.
>
> (Plunkett Foundation representative)

In conventional food systems, buyers/customers weigh up economic reasons to decide whether to exit a trading relationship, normally doing so in an indirect and impersonal way (i.e. not picking a product from a supermarket shelf or a catalogue in the case of buyers). In MSCs, this model is turned on its head, prioritising instead the option of voice over exit (exiting the trade relationship; Hirschman, 1970); by institutionalising the voicing of concerns and disagreements, economics become politics and social relations, and the indirect, impersonal approach of supermarkets becomes direct and messy. Trade relations become more personal, identities are known, and interdependence is not only acknowledged in principle but also in the governance of the cooperatives through weighted voting. They also remove distant, anonymous shareholders from the equation, as members are local to the cooperatives.

The model involves a certain degree of contact between farmers and buyers, which is considered a strength as it increases their resilience, but also a challenge in terms of time and geographical constraints. If it continues to grow, MVP has already identified a federated model of sister cooperatives as an adequate strategy for expanding without losing close working connections. But, already, one of its challenges is to find ways to get all members at the same table, a hard task taking into account buyers and growers have virtually opposite timetables. As one of the MVP members reflects: "[some of the buyers] don't have the time or as much inclination really to get involved I think", but it is acknowledged that it is not just a matter of timetables, but also ways of working, as this quote reveals:

> It's a very new thing for these businesses, new in lots of ways, firstly in how they buy from MVP, they're used to just phoning up at 11 o'clock at night at the end of their service and ask for produce that would be delivered first thing the next morning. With MVP they have to order more

in advance because we pick to order. Secondly, they're not used to any fruit and vegetable supplier asking them to be involved in the business and thirdly none of the restaurants or the university were used to cooperative work. So it's new in many levels, I think quite challenging, so if their participation hasn't been as great, I think there are loads of good reasons for it.

(MVP member)

The above comment reflects the complexity of bringing together diverse memberships with, not only different perspectives and interests, but also different routines and timetables, which some authors believe will take MSCs down the route of conversion (Lindsay and Hems, 2004). However, this can also be seen as a first step and part of the mission of MSCs: coordinating and bringing together unlikely allies in the pursuit of common needs. Authors agreeing with this view suggest that having a long-term view is key, and that, with time, rather than risk of conversion, what emerge are reduced transaction costs thanks to the multi-stakeholder approach that structurally fosters communication, trust and engagement (Lund, 2012).

At the same time, having direct interaction with other groups of actors increases accountability, as growers and producers know they are going to see each other regularly. This regular contact is also part of trying to introduce a variety of methods of governance, beyond attendance at decision-making meetings – for example, involving MVP buyers in choosing varieties grown by farmer members. Reconnecting with different groups also utilises the collective intelligence and knowledge of the membership, sometimes creating significant savings, as in the case of one of Esnetik's members who shared €10,000 worth of product development for a local version of a Camembert cheese with fellow members (Ajates Gonzalez, 2017d).

ACs have evolved in different ways from other practical manifestations of the cooperative movement, especially in comparison with workers' cooperatives. Perhaps the fact that members of workers' cooperatives have to work together every day makes a difference to the way their relationships develop. Farmers do not have to cooperate daily; in fact, it could be said that they hardly ever cooperate, because, as Wilson and MacLean have described, farmers have delegated the act of cooperating to professional managers (Wilson and MacLean, 2012). The reasons for joining a workers' cooperative can also be different from the outset from those that farmers might have for joining an agricultural cooperative. Farmers have a passive rather than productive relationship with their cooperatives, mainly attending meetings and most being distant from the governance of their cooperatives, which often have a similar managerial hierarchy to that of private companies. Workers are the most critical part of the business in other cooperative sectors, but, in farming, resource-sharing and bulk buying seem to be the key elements. This is reflected in the nearly non-existent number of workers' cooperatives in agriculture. OrganicLea is an exception to this rule. When asked how it compared with conventional ACs, one of OrganicLea's members answered:

> There are clearly many differences, lots of differences, but also many simi-
> larities. It means we have more in common with other worker cooperatives
> I suppose. But it's different because we are working together and making a
> lot of decisions together on a day-to-day basis, whereas in that kind of more
> classic AC you're coming together for buying or you're coming together
> for selling or you're sharing equipment but you're not really sharing that
> day-to-day decision-making [. . .] it's hard to compare, but definitely the
> workers' cooperative element adds another layer of working together. But
> also that means that cooperative relations are the total basis of the organisa-
> tion, that everyone has an equal voice, everybody has an equal salary.
>
> (OrganicLea member)

This quote reflects an attempt to construct less hierarchical organisations and a
more horizontal conceptualisation of leadership, a reflection of the prefigura-
tive politics of the cooperative models participants wish to develop (Sutherland
et al., 2014). This aspiration to create "leaderless" social movements can be
seen as their attempt to prove how the democratic and horizontal society they
wish to see is not only possible but also feasible (Sutherland et al., 2014).
Consensus decision-making is time-consuming and messy, but it is considered by these MSCs to be an indispensable element of participative economic
democracies of food production and provision.

Co-producing commons

From the environmental commons of soil, water, air, nutrients and energy
used to produce food, to ecological public health (Rayner and Lang, 2012),
no other topic brings together so many aspects of common goods as food does
(Ostrom, 1990 and 2009; Böhm et al., 2014). In this section, the concepts of
local food and environmentally friendly farming are used to discuss how these
MSCs' practices relate to the commons.

All the case studies have emerged on the edges of, and are dependent on,
cities, acting as a link between urban and rural actors. When unpicking the
reasons given for the geographical delimitation imposed for the produce traded
by these cooperatives, short distances come up as intrinsically connected to
discourses of provenance, sustainability (reducing food miles) and quality
(freshness), but also as highlighting the value of face-to-face contact between
members. Localness is a cross-cutting theme that acts a normative but also
identity aspect of these MSCs. At the same time, it uncovers tensions in relation to urban consumers' wants, producers' wages and fairness. Local food is
perceived as a common good that needs to be collectively maintained, con-
fronting a reality that shows consumers want it but do not want to pay for it:

> [E]verybody is talking about wanting local, sustainable food, and "we want
> more local food", so then you have to bloody pay for it, because you can't
> expect people to go into a back-breaking, high risk job, you know?
>
> (MVP member)

Their aim is to produce environmental commons through organic methods and financial and social commons through a stronger local economy. And linked to local food is the concept of seasonality. A common criticism of narrow localist approaches highlights the fact that local food does not necessarily have a lower environmental footprint if grown out of season (Sharzer, 2012; Baggini, 2014). Both MSCs aim to educate members about seasonality as part of their mission, a topic especially relevant to MVP's work around supporting members to make their menus more seasonal (also a key strategy for reducing costs, as produce is cheaper when in season, opening potential scaling-up routes through public procurement by fitting into their existing tight budgets).

Esnetik's political-economic conception of organic production weaves together the concepts of open knowledge transfer and sustainability:

> The problem is that being an organic farmer is much more complex and demands much more training than a conventional grower, why? Because a long trajectory is needed, a lot of experience, whereas in the other agriculture, you are given everything done. When you have a pest problem, you go to the nearest all-too-typical "pharmacy" and they give you the product. Here, people who have the experience are the ones who are going to transfer how to act against pests, how to treat the soil.
>
> (Esnetik member)

Organic agriculture as a result becomes a political act of autonomy, both as an approach to production but also with regard to knowledge acquisition. These horizontal knowledge exchanges can be seen to serve three roles: on the one hand, they require farmers to be proactive, shaking them off the spoon-fed dependency spread by industrial farming; on the other hand, those processes of knowledge transfer strengthen farmer networks and interaction as well as collective knowledge in the cooperative. Finally, by fostering informal processes, rather than standards-based approaches for certification or labels, Esnetik reduces the risk of "conventionalisation" as discussed by Goodman in relation to the new depoliticised versions of the organic and fair trade movements that have been absorbed by large processors and retailers in the food system, where the organic aspect is treated just as any other selling point (Goodman et al., 2011). By engaging civil society organisations on its board and in its decision-making, Esnetik has identified both a barrier to producing commons and a way to overcome it:

> [T]hat is it in one word; that is it, to get them to roll up their sleeves. This is hard, you know? It is hard because they are very theoretical on their foundations, even in the area of consumption, it is hard because consumption has been much more theorised than production.
>
> (Esnetik member)

This analysis of the emphasis on locality in the case studies has revealed how the rural–urban divide is not considered in the Open Coop framework, but is key to food studies. Cosmopolitan localism (Morgan and Sonnino, 2010), a concept discussed in the next section, brings together the requirements of the Open Coop framework by calling for the creation of strong, healthier local communities and economies with a global awareness and participation that look after their environmental commons (Manchester Veg People, 2016).

Global outlook: socialising knowledge

Cooperatives have been described as potential leading actors in shortened food chains as they make possible "economies of scope" or "synergy" versus simplistic "economies of scale" (Marsden et al., 2002). Pratt and Luetchford have highlighted how political positions and political aims can act as a common ground where both consumers and producers with apparently clashing interests can meet and strive (Pratt and Luetchford, 2014:181). In this context, this section discusses how the case studies bring political economies of scope and synergy to local food systems.

Evidence from the case studies reflects how MSCs can bring together economic benefits from collective bargaining while also serving as a melting pot for environmental, political, health and livelihood concerns. Their wide range of strategies to attempt change comes from their long-term vision down to their day-to-day practices. Starting with their democratic governance model, the efforts focus on re-localising trade, knowledge and financial returns as well as production and consumption in an increasingly global food market.

The uniqueness of their produce comes from how it is grown and traded, giving them a comparative advantage over competitors who cannot guarantee either the organic and local attributes or the values and practices associated with their production, transforming the offer in their local areas. These MSCs are also transforming purchasing practices, moving them away from inertia by sharing knowledge about food issues with members and asking them to get involved in collaborative crop planning, farming activities or adapting menus to seasonal ingredients.

Despite their focus on locality, cooperative members interviewed for this study were aware of the dangers of short-sighted "defensive localism". This term has been defined by Morgan and Sonnino as a "narrow, self-referential and exclusive" alternative to the conventional food system. Instead, these cooperatives have opted for what the authors refer to as "cosmopolitan localism", which in contrast is "capacious, multi-cultural and inclusive" (Morgan and Sonnino, 2010; Marsden and Sonnino, 2012). Apart from being formed of heterogeneous memberships, these MSCs are also part of wider networks, connecting with other players at local, regional, national and international level and being active and aware of the "global struggles" striving for a better food system and a fairer economy (for evidence of involvement in wider networks and projects, see the following references: Kindling Trust, 2012; Actyva, 2015; CASI, 2015; Esnetik, 2015; Land Workers' Alliance, 2015). These practices are partly in line with the last condition for Open Coops with regard to socialising their knowledge, but work to create closer connections among other cooperatives needs to be developed. Some efforts in this direction have started: for example, Actyva regularly attends national meetings of MSCs in Spain; MVP and OrganicLea are collaborating in rolling out their FarmStart initiative; Esnetik is exploring a potential agreement with a cooperative of drivers to reach more consumers and overcome logistic barriers, and so on.

As opposed to other actors in the food system who are concerned with offering a unique service to retain a comparative advantage, MSCs *want to be copied* (but not co-opted) and make an effort to share their experiences and business plans to encourage a localised reproduction of their models (Kneafsey, 2015).

Being part of wider national and international networks, (e.g. the Solidarity Economy European Network, Via Campesina, Oxford Real Farming Conference or the Land Workers' Alliance), these MSCs are socialising their knowledge and practices with other groups working on related issues (alternative currencies, food sovereignty movement, etc.). For example, MVP's work is "inspired and guided by a radical perspective that identifies the need for significant social change" (Manchester Veg People, 2016). MVP has also openly shared its original business plan to encourage the growth of other MSCs. MVP operates an open-book policy and shares its accounts with any member who has been a member for 6 months. This emphasis on openness and transparency is also followed by Actyva, reflected for example in its willingness to publish its performance online and share the learning from its first months and initiatives.

Through these networks, they are also building a sense of interdependency with other, weaker groups of actors in the food system (e.g. the individual consumer), but also with other sectors of the (solidarity) economy. These wider, looser networks interlink "multiple, rhizomatic grassroots movements" (Böhm, 2006:167), which enables them to gain strength and share ideas in order to increase their impact, but is also their way of reflecting their values in their activities.

Another example is Esnetik's vision for a "feminist, solidary and sustainable economy" (Esnetik member). For Esnetik, reaching out to NGOs and local authorities is a way of expanding its impact, not a sign of weakness. For its producers, to be organic means to be autonomous. Autonomy is not understood as unconnected independence; the members' sense of urgency and real awareness that they must become allies of consumers and the environment have shaped what could be termed as an "autonomous interdependence" model of interrelations and dependencies among producers, workers and consumers, on their own terms.

This is in line with Böhm and colleagues' analysis of grassroots organisations, in which the project of autonomy is seen as essentially collective (Böhm et al., 2010). For the authors, autonomy cannot be completely fulfilled, as there is a constant risk of being co-opted by dominant political and financial regimes, creating an ongoing but fruitful tension:

[A]utonomous practices are rarely completely captured by existing institutions. This means they continue to produce the possibility of resistance and change. Thus, we want to conceptualise autonomy neither as a positive force of unconstrained creativity nor as part of the ongoing movement of negative dialectics. Instead, we rather emphasise the antagonistic tension between positive forces of creation and negative dialectical challenge involved with autonomy.

(Böhm et al., 2010:27)

Debates and struggles about autonomy generate contested meanings and practices, creating opportunities for change (Böhm et al., 2010). In the case of farming, Patrick Mooney saw this potential of ACs as organisations that can legitimise and sustain class struggles, tackle power imbalances and improve workers' conditions (Mooney, 2004). In this sense, Mooney noticed that ACs provide the space for those tensions to emerge, become visible and provide innovative solutions, in contrast to the neoclassical economics model that presents those tensions as barriers to profit-making efficiency (Mooney, 2004).

Münkner pointed out that the MSC is not a brand new concept, but "it corresponds to the original mission of cooperatives to render services in all aspects of life" (Münkner, 2004). In this sense, MSCs are also challenging modern models of cooperation, especially Actyva, which offers other services beyond food. As one of its members put it:

This is not a new approach, but a return to the original vision of the movement; the cooperative models of the 19th century [. . .] were very inclusive, all facets of human life; maybe this aspect has been a bit relegated in the 20th century but it is being incorporated again.

(Actyva member)

Another effort to create a wider impact is their work to change public procurement, which has been identified as a key route to a fairer and greener food system (De Schutter, 2014). Actyva is considering this route, and MVP is already tapping into this scaling-up strategy, spreading the common good of local, fresh organic food to a wider range of consumers. MVP values public procurement collaborations as it sees them as a way of democratising locally produced organic food and an opportunity to scale up its initiative in order to create more jobs in farming.

By aiming for a multidimensional vision of sustainability, MSCs can attract the interest of those interested in specific aspects – for example, organic or local or fair pay – educating them later about existing interconnections in the food system. As highlighted by Mooney, "in the realm of social relations, cooperatives provide an interesting site for the exploration of tensions [. . .] on the social relations of production and social relations of consumption" (Mooney, 2004:80–81). Furthermore, they act as urban–rural links, and their networks with civil society organisations go from local to global. Nevertheless, local initiatives cannot achieve change unless replicated at a global level.

Legal aspects

During the thematic analysis to investigate the dynamics and experiences of these four MSCs, the theme of cooperative legislation emerged several times, and it is discussed here as it adds an extra layer to the Open Coop framework.

In 2009, the Somerset Cooperative Rules were created in the UK to specifically support MSCs wishing to weight the votes of their different member types (Somerset Cooperative Services, 2009). MVP welcomed the new Somerset Rules for MSCs, but the model is still very new, and members, especially those not familiar with cooperative working, have to be introduced to it, which requires time and resources. The model helps the cooperative to avoid replicating the system they are trying to change. These MSCs remind us how legal frameworks often lag behind social practices.

There is a historical lack of clarity about what type of cooperative is best for rural or urban contexts, as a cooperative worker from the cooperative development agency that created the Somerset Rules explains:

> [A]griculture poses some particular challenges for cooperation, it tends to involve people spread out over quite a large area, maybe working most of the time by themselves or very small groups, so perhaps it is not immediately straight forward to create agricultural cooperatives. And it's also I think the point with the two great traditions of cooperation: workers cooperation and consumer cooperation, sort of collide, but . . . there's a general assumption that in retail one expects consumer cooperatives and in manufacturing one expects worker cooperatives but in agriculture it's not altogether clear and people have tried different approaches and I think what we decided to do in SCS was to say, well, maybe both stakeholder groups are of more or less equal importance and we want to find somehow a way of balancing the voice of the large number of consumers with the voice of the smaller number of producers, and so that was something that was very much on our minds.
>
> (SCS representative)

This section has highlighted the ongoing issue of how co-operatives are still part of a market economy and need to fit in with existing legal frameworks or struggle to create their own. The evidence from the case studies suggests that having an adequate legal framework is not sufficient to encourage MS initiatives owing to the bottom–up nature of these associations, but it can create the right policy context for MSCs to flourish. The MSC legal model in the UK, with weighted voting, although still very new, seems like a positive development. More support is needed to advise interested groups on selecting and applying the right legal frameworks for their specific case. At the moment, as both case studies suggest, this aspect is absorbing time and resources that could be invested in other activities, such as marketing and engagement with members.

Six more-than-economic benefits of MSCs

The case studies suggest MSCs operating in food and agriculture have the potential to generate a diverse range of more-than-economic benefits associated with their cooperative structure and operations. These potential benefits emerge when actors are brought together who are used to thinking they have opposite interests, and by questioning the shortcomings of the hegemonic food system, challenging and competing with long-established cultural, legal, dominant cooperative imaginaries and economic norms. The opportunity of the benefits being realised thus often encounters difficulties, creating a push–pull dynamic characteristic of initiatives that are operating in a specific socio-economic context that they are trying to transform at the same time. These benefits and counter-arguments can be classified as follows:

Economic

Although all the case studies are struggling to achieve financial sustainability, their abundant social capital has helped them compensate to a degree for the lack of financial capital. Despite the small scale of their operations, the fact they exist is sufficient (much like the organic or fair trade movement) to create debate, help raise standards, keep mainstream retailers on their toes (Sanchez Bajo and Roelants, 2011) and keep the ongoing "rural struggle" alive (Mooney, 2004). However, currently, the price of food items is often not related to the cost of production, as retailers and subsidies distort this relationship. A formula for pricing is still missing; fair pricing and true cost accounting for food and farming are still an unsolved problem (Sustainable Food Trust, 2017). Produce from MSCs is perceived as expensive, as it incorporates externalities of food production (but less waste means the overall price per unit can be the same). Often, MSCs cannot be convenient or efficient in the short term; they need a long-term view, but exist in market economies guided by short-termism.

Sectorial

At the same time, MSCs confuse the simplicity of the old reductionist agricultural paradigm characterised by its clearly divided market roles. The integration of different actors involves a process of negotiating expectations and clashing interests in order to achieve middle ground, especially around the uncomfortable topic of agreeing a "fair price"; they attempt to do so by prioritising voice over exit mechanisms, which is time- and energy-consuming but can be a resilient and sustainable governance approach in the long term (Lund, 2012).

Additionally, the dominant reductionist paradigm of agriculture has invested a lot of effort and capital to transform peasant growers into business entrepreneurs and change the rural image of farming in order to project a more businesslike one (Lang et al., 2009). The "get big or get out" motto of conventional agribusinesses, aiming for more mechanisation and less reliance on human labour, is, however, not necessarily a synonym of success for these

MSCs, especially because one of their objectives is to create more employment opportunities in farming, not fewer. In this sense, these MSCs are challenging mainstream meanings of efficiency and indicators of success. Intrinsic to their MS structure, there is also a challenge to conventional definitions of "member" in agricultural cooperatives in their opening their doors to consumers and workers. But their resistance goes beyond membership. Agricultural cooperatives are also pushed to get big or get out (Gray and Stevenson, 2008); in that context, these MSCs are proposing a post-productionist version that goes beyond the bulk savings and dividends focus of conventional farming cooperatives.

Legal

Their new legal form shows that a different way of doing things is possible, but it is very resource-demanding. However, the MSC legal model is new and complex and relies on the willingness of members and forward-thinking business advisors to propose it to new enterprises as it is not very practical or convenient to get established. On the other hand, the MSC legal form is a positive development as it allows members to have a voting system that does not replicate the power imbalances they set out to eliminate in the first place.

Cultural

The dominant culture of consumer choice is one of the pillars of the conventional food system that is confronted by the MSC model. By mainly offering crops that can be grown locally, these MSCs are challenging conventional understandings of progress based on ample choice. However, at the same time, they are offering choice of new varieties and new local products not offered by large retailers. They are also maintaining the local food growing culture by supporting new entries into farming.

Policy and public procurement

MSCs highlight the unfavourable policy contexts for both subsidies and public procurement contracts, while also providing a route to challenge and improve current practices. In the case of MVP – and in the future maybe Actyva too, if its members decide to go down that route – tapping into and transforming public procurement's easy and long-established contracts, which seek value for money at the expense of quality and fair returns to producers, are both part of its mission and its income strategy. Finding public organisations that have a champion with a long-term vision and willingness to invest the time to explore how to navigate contract requirements and test new seasonal menus that still match their budgets is not an easy task. However, the benefits of tapping into public procurement are huge, both for MSCs, as it gives them financial stability, and also for the general public, as it democratises access to local and organic food.

Despite their local focus, MSCs are – at a local level and a national level through their networks – contributing to the debate on self-reliance versus global food trade promoted by policies formulated by the European Union and the World Trade Organization (Lang et al., 2009). These MSCs do not exist in isolation with the freedom to shape trade to their vision, but in a complex policy arena of multilevel governance.

Academic

MSCs pose a challenge to academia's often divisive disciplinary lenses. As Goodman and DuPuis (2002) have highlighted, agricultural economics is separate from cultural theories of consumption, and this division needs to be overcome. More than a decade ago, Mooney called for a new sociology of cooperation in production and consumption (Mooney, 2004); MSCs revive that call and push food scholars to develop this line of research and acknowledge and analyse real experiments happening on the ground.

The interconnections of the above categories are clear. The multifaceted character of MSCs as enterprises, social networks, objects of studies and legal entities allow them to act as a lever for change in different facets of society.

Conclusion

This chapter started by reviewing historical attempts to create MSC initiatives in food and farming and then discussed the current literature on MSCs. The evidence presented shows how the introduction of different types of member seems to both complicate and enrich the cooperative mission, both theoretically and in practice (Moog et al., 2015). The findings suggest that cooperatives in the food sector, striving to go beyond economic benefits, are moving into the arena of the solidarity economy and linking up with other global social movements. Often, their endeavour to aspire to multidimensional definitions of food sustainability hinders their financial sustainability. Other economic, policy, cultural and sectorial barriers that preclude MSCs' potential to create systemic change were discussed in relation to Bauwens and Kostakis' (2014) Open Cooperative framework. This framework was applied to four case studies of MSCs, based in Spain and the UK. The framework calls for a model of open MSCs that meet four requirements: dealing with negative externalities, participatory democracy, co-production of commons and global outlook.

At a time when capitalism is concentrating not only capital but also its workforce (namely consumers) in cities, MSCs represent an innovative model with the potential to decentralise power in the food system – for example, diffusing supermarkets' control of food sales. However, the MSC model requires robust efforts from both ends of the food chain: farmers going downstream and consumers and buyers going upstream, adopting more relational ways of trading, based on geographical and temporal connections. The findings from the four case studies presented show that the MSC model is demanding and messy, reflecting a genuine and passionate attempt to realise all the criteria of multidimensional sustainability and linking environmental and health concerns with a call for social justice (Tornaghi, 2014; Kneafsey et al., 2016). Being agents of transformation, MSCs' practices are both oppositional and alternative, putting into practice politics of collective responsibility.

The MSCs discussed are aspiring to transform the food system by following two strategies: first, by raising awareness of and reducing unaccounted externalities of food production; second, by aiming for multidimensional sustainability, creating a wider campaign space where efforts to address the new fundamentals can come together. Ostensibly, they seem to have unclear objectives with regard to co-production of commons: are they about realising a closed local economy? Championing greener methods of production? Or providing affordable food? Creating farming jobs and getting a good return for producers? Unlike the "monoculture cooperativism" that only covers one aspect of members' lives, MSCs exemplify a much more diverse and diffused notion of cooperation, the multiplicity of their objectives being a reflection of the increasing policy stretching that global food governance is experiencing.

At the organisational level, if principles, governance and outward networks are cultivated, MSCs can become a connecting link between bottom–up initiatives and top–down food policies, a link with the potential to be scaled up by collaborative public procurement strategies. Their heterogeneous memberships and networks make them complex but also grant them resilience, contacts and a voice on different platforms of the multilevel governance of global food policy.

The policy implications of the findings will be discussed in detail in Chapter 10; however, it is interesting to highlight here that the findings suggest an adequate legal framework is not sufficient to encourage MS initiatives owing to the bottom–up nature of these associations. Nevertheless, a suitable legal framework can be part of creating the right policy context and can minimise the diversion of much needed cooperative resources. This point also raises questions with regard to current farming subsidies for rural collaboration and whether these funds might be better invested in supporting new models of MS cooperation that foster links between rural and urban development. EU policymakers must consider what kind of cooperation they want to reproduce through subsidy support. However, the challenges facing MSCs are related, not only to multilevel governance, but also to long-established cultural and social norms and expectations around choice, price and progress that affect perceptions of farming and buying habits, both at the domestic level and in public procurement.

Note

1 Making Local Food Work was a £10 million programme that ran from 2007 to 2012 to support the general public to take ownership of their food and where it comes from by providing advice and support to community food enterprises across England.

9 Third spaces

Fighting the cooperative corner and interrogating the alterity of emerging cooperative models

Chapter 8 has discussed how in an AC sector dominated by competition, vertical growth and focus on markets rather than members, some new emerging alternatives in farming still look at the cooperative form as a useful mechanism to realise their objectives, instead of as an outdated model to distance themselves from. This chapter offers a deeper examination of the practices these alternative AC models use to reduce the risk of co-optation by dominant trade dynamics of powerful players.

Alternative food networks are often presented in the literature as contrarians and outside the industrial food system. However, Goodman et al. (2011) have pointed out how many AFNs are presented as "oppositional" even when they still rely on capitalist market relations and/or the state for their reproduction. In some cases, the success of some AFNs becomes their failure when growth means being incorporated into the dynamics of industrial models they were born to replace in the first place.

When the strategies of AFNs are co-opted by the globalised industrial food system, their transformative power is reduced or neutralised. Example of this trend is the organic movement that first started as a grassroots initiative and gradually became absorbed by large retailers as a mere additional product line (Buck et al., 1997), similarly to "local" (but not necessarily ethically or sustainably grown) and fair trade foods (Taylor, 2005; BTC, 2014).

I have argued somewhere else that one of the biggest challenges facing the food system is in fact the continuous co-optation of potential solutions by the dominant food regime (Ajates Gonzalez, 2017b). This process perpetuates current dynamics and suffocates more balanced alternatives. By applying the concept of third space as an analytical lens, this chapter tries to go beyond arguing whether MSCs' initiatives are indeed fully embedded in market relations or not, and focus instead on how and if they are able to maintain their alterity and, more importantly, in what areas new spaces for developing and maintaining non-co-optable practices can emerge.

With the above aims, the chapter will offer a dual contribution to the AFN literature. First, it puts forward the argument that a new wave of emerging niche cooperatives are opening up "spaces of possibility" in opposition to the dominant agri-food regime (Holloway et al., 2010; Goodman et al., 2011;

Ajates Gonzalez, 2017b). Building on Goodman et al.'s critical analysis of how AFNs have to navigate their coexistence in the mainstream circuits of capitalism, with "its own dangers and contradictions – discursive, material and political" (Goodman et al., 2011:245), the chapter explores how these tensions unfold for six small cooperatives.

Second, the chapter builds on the counter cooperative-degeneration argument proposed by Arthur et al. (2008) by providing supporting evidence that backs up the validity of their concepts of deviant mainstreaming and incremental radicalism. These terms attempt to capture the internal dynamics of social spaces trying to remain deviant and sustain a degree of alterity while remaining financially viable (Arthur et al., 2008).

Chapters 6 and 7 revealed a growing trend of co-optation of ACs that are adapting to, instead of challenging, dominant food system practices. In this context, to what extent is the legal model of cooperatives still relevant for innovative experiments in farming? Are the ICA cooperative principles still significant for emerging cooperative models bringing together consumers and producers? What strategies do alternative ACs use to remain alternative in their practices and reduce the level of co-optation by the industrial food system?

Third spaces, deviant mainstreaming and valuing the marginal

The Wales Institute for Research into Cooperatives (WIRC) proposed the concept of *deviant mainstreaming* after studying cooperatives for seven years with the aim of capturing the internal dynamics of autonomous transformative social "spaces" that try to remain "deviant" and sustain a degree of alterity while surviving in the dominant system, that is, neo-liberal capitalism (Arthur et al., 2008). In turn, the concept of *incremental radicalism* weaves together the process and outcome of organisations applying deviant mainstreaming; it refers to the organisations' capacity to inspire others, resulting in more transformative spaces becoming "emancipated" and creating the conditions for being "more widely challenging of the processes of domination" (Arthur et al., 2008:31).

Building on these counter cooperative-degeneration arguments and concepts put forward by WIRC, this chapter draws on Homi Bhabha's anthropological notion of "third space" to unravel and frame the social innovations that are emerging out of shared *deviant* food spaces, both physical and ideological, and between rural and urban spaces and people.

Chapters 6–8 revealed a wide range of ACs and diverse practices. Chapter 8 provided evidence of how some new ACs are attempting to colonise the spaces between the more corporate cooperatives at one end and the more radical, non-legally incorporated and more informal initiatives at the other end of a continuum that includes many hybrids in between. Human practices existing in between two very different spaces of praxis are not new phenomena, having been theorised long ago in anthropological studies.

The concept of *third space* was developed in anthropological studies by post-colonial writer Homi Bhabha (1994). Bhabha's concern with colonisation made him aware of the overlapping spaces that some native people found themselves in, new spaces that combined features of their old culture and that imposed by the colonisers. These hybrid, or third, spaces were contested territories in which new identities, languages and cultural practices emerged, characterised by an "unpredictable and changing combination of attributes of each of the two bordering spaces" (Muller, 2007:53).

Bhabha's theory of cultural hybridity opens up a window into these fertile environments for innovation in food and farming, these "in-between" spaces that "provide the terrain for elaborating strategies of selfhood – singular or communal – that initiate new signs of identity, and innovative sites of collaboration, and contestation, in the act of defining the idea of society itself" (Bhabha, 1994:2). It is argued that, because of this contestation, this questioning and the arising challenging situations, room for reinterpretation and renegotiation opens up. Being in this in-between space, reality and temporality fosters creativity, as "cultural production is always more productive where it is most ambivalent and transgressive" (Bhabha, 1994:2). This concept helps to analyse the dynamics of these new cooperative initiatives emerging at the interstices of well-established food cultures and regimes (Friedmann, 2005; van der Ploeg, 2008) and to unravel the new identities that farmers, activists and consumers experience in multi-stakeholder cooperative models as well as the new sets of practices, language and values that are created during these interactions.

A related principle that has strong parallelism with the concept of third space and is common within alternative food growing practices is that of "using the *edge* and *valuing the margins*" promoted by permaculture. Permaculture is a design approach popularised by Bill Mollison and David Holmgren in the late 1970s, often land based and related to food growing, but increasingly applied to other types of project or dimension, such as furniture design, relationship with time, money or well-being (Mollison and Slay, 1991; Social Landscapes, 2018). Permaculture has a fixed set of ethics and principles that are based on interconnections and interdependence of elements in natural ecosystems (Garnett and Godfray, 2012). One of its 12 design principles is "use edges and value the marginal", based on the idea that productivity and number of species are higher in shared edges or boundaries where two ecosystems come together, as the resources from both sides are available to support life (Peeters, 2011).

This concept serves as a useful metaphor in food and farming studies research by emphasising the value of often unheard voices, such as those of small farmers, women farmers, consumers on low incomes and those moving between dichotomies often used to divide the current food system and its actors: the rural and the urban, consumer and producer, legal and illegal ways of producing and distributing food (e.g. raw milk) and even keeping seeds. When applied to the MSC ecosystem we will be analysing next, the edge and third spaces created between cities and rural areas, between urbanites and *ruralites* and the marginal – from people at risk of social and financial exclusion to disappearing breeds and varieties – become active and useful parts of the whole.

Third spaces in the literature, food and retail

Bhabha's concept has become highly influential and has transcended disciplinary boundaries: Bachmann-Medick (1996) applied the concept to translation theory; Soja (1999) proposed a theory of third space to reconceptualise space and spatiality. The term has also been used in planning, education and linguistic studies (Grenfell, 1998; Gutierrez et al., 1999; Moje et al., 2004). When applied to information technology participatory design, the benefits of fostering third spaces (in between the user's domain and the technology developer's domain) have been described as creating the conditions for "challenging assumptions, learning reciprocally, creating new ideas, which emerge through negotiation and co-creation of identities, working languages, understandings, and relationships, and polyvocal (many-voiced) discussions across and through differences" (Muller, 2007:166). As described in the next section, similar benefits emerge in the case studies. Table 9.1 (adapted from Muller, 2007) offers a summary of Muller's list of claims related to third spaces in relation to the practices of the cases presented.

In retail, the idea of third spaces has been increasingly appropriated by outlets, especially clothes shop chains, in an attempt to fight digital commerce by focusing on providing "experiences" to customers and creating spaces that encourage shoppers to visit retail spaces (JWT Intelligence, 2014; Thompson, 2014). In food, venues such as the Third Space Café in Dublin have emerged in the shape of a social business venture that describes third spaces on its website as "neighbourhood places where people can gather regularly, easily, informally and inexpensively" (Third Space Café, 2012). Bigger players have also entered the third territory: Starbucks has been positioning itself for over a decade as a third space in between home and the office, a place to relax, but also to work and hold meetings (Schmitt, 2003).

At the other end of the supply chain, third spaces around food production have also emerged with the rise of urban agriculture, creating new demands and new challenges for urban planning and municipal regulations. Front gardens and public areas become symbolic spaces, often creating conflict between conceptions of "culturally appropriate" and "legitimate" use of space that stir debate (Schindler, 2012). In this context, efforts by municipalities to separate agriculture from residential uses are interrogated, and the question of what land is for arises.

The actual label of "third space" has rarely been applied to describe people; however, it has been used in the media to refer to the growers of Los Angeles's South Central Farm. This farm was considered at the beginning of the current century, the largest urban farm in the US, cultivated by urban dwellers, mostly of Latin American descent. During their long struggle to remain on the land, which resulted in eviction, they received much media attention (including the documentary film "The Garden") and were described as "third space farmers", as their activities were compatible with the needs of the community and had a combination of "human and natural capital" (Katz, 2006). The farm had a high number of plant species and varieties that were considered and used by their

Table 9.1 Summary of claims related to third spaces

Characteristics of third spaces	Examples of practices from case studies that adhere to these claims
Overlap between 2 (or more) different fields (inbetweeness)	• Spatially: the case studies operate in between urban and rural spaces
Marginal to reference fields	• Niche, not perceived as a threat to dominant system
Novel to reference fields	• Novel MSC models • Innovative double labelling practices • New inter/national networks being developed
Not "owned" by any reference field	• Reducing dependency on agri-business by aiming for closed farming systems • Some are dependent on external funding but are in a period of transition towards being financially independent
Partaking of selected attributes of reference fields	• Knowledge of actors from different fields (where they still operate, both consumer and producers): e.g. experience around navigating the issues of logistics for transporting produce • Knowledge of local and export markets • Previous experience working in cooperative enterprise embedded in long supply chains for exports
Potential site of conflicts between/among reference fields	• Difficulties with the administration regarding not-for-profit status • Taking share of local market for local organic produce • Raising expectations from consumers, e.g. expecting local food in their supermarkets • Raising expectations from buyers: e.g. expecting local organic food from providers but also to have a say on what's grown and how
Questioning and challenging of assumptions	• Challenging assumptions: What is local? What does expensive mean? Is price related to unit or related to overall usable amount? What is land for? What is agriculture for? Speculating or making a living from food production? Maintaining cultural traditions? What does the typical farmer look and sound like? • Challenging the assumptions that (1) farmers and consumers have unreconciliable objectives; 2) people at risk of social or financial exclusion are of no value to capitalist societies • Challenging gender stereotypes in farming
Mutual learning	• Between growers, between urbanites and ruralites • Events with workshops for members educating about traditional varieties, crafts, recipes, educational events in schools
Synthesis of new ideas	• Co-defining relocalisation, quality, terms and conditions, e.g. FairCoop

Negotiation and co-creation of:

Identities	• Such as: "urbanites in transition"; multi-stakeholderism; integral cooperative member; identity of the MSC itself within the cooperative sector/movement
Working language	• Creating new terms such as "urbanites in transition", "neo-rurals"
	• Changing words to aspire to a non-gendered discourse
Working assumption and dynamics	• Frequent discussion and meetings
	• Participation, social media platforms
	• Consensus decision-making
Understandings	• Food as a right
	• Acknowledging and valuing interdependence among humans and nature
Relationships	• Based on principles of social justice, mutual aid, solidarity
	• Trade relationships change, e.g. alternative currencies
	• Relationships with other organisations and movements working to change the food system
Collective actions	• Frequent, celebratory education and training, open days
	• Multilevel – to influence policymakers (e.g. MVP – leading on Manchester's food strategy)

Dialogues across and within differences

Polyvocality	• Diverse multi-stakeholder membership from different backgrounds and with different objectives
What is considered to be knowledge?	• Stretching the meaning of data, valuing all knowledge
	• Knowledge is shared between producers, other members and other organisations: e.g. during events, Esnetik's recipes, MSCs congresses, etc.
What are the rules of evidence?	• Evidence comes from all directions, top–down from trusted sources and bottom–up through members; e.g. market info, performance of varieties
How are decisions made?	• Consensus decision-making
	• Weighted voting
	• Informal and then formal discussion
Reduced emphasis on authority – increased emphasis on interpretation	• Diverse cosmovisions from diverse memberships
	• Shared authority and decision-making powers
Reduced emphasis on individualism – increased emphasis on collectivism	• Cooperative working
	• Realisation of others' needs and strengths by incorporating them as members of the cooperatives
Heterogeneity as the norm	• Diverse multi-stakeholder membership
	• Diverse crops
	• Several cooperative activities and financial strategies to survive
	• Valuing the marginal (e.g. small producers) as a core principle

Source: Adapted from Muller, 2007

growers, not only as food, as conventional commercial farmers would, but also as medicine, for spiritual purposes, for infrastructure (for example, using cactus patches as natural fencing and food) and companion plants (Peña and Foucault, 2005). By bringing plants and a variety of uses from their cultural traditions, these farmers were making place and making home away from home.

However, despite the explicit reference to the term "third space" in the above examples, their conception of third spaces is quite narrow, both spatially (with a clear focus on urban areas) and demographically (organised by fairly homogeneous groups of people); apart from the South Central Farm case, these initiatives do not attempt to question and transform the wider systems they exist in.

Opening up spaces of possibilities and closing gaps to maintain resistance

MSCs are cooperatives with two or more classes of member; for example, in food, MSCs can have a membership comprised of consumers, producers, buyers and workers (Ajates Gonzalez, 2017a). The governance becomes more complex, but it acknowledges the interdependence of different actors in the food system. They challenge the often-held assumption that consumers and producers have irreconcilable aims. The transforming potential of the model has been recognised by many civil society actors, including anti-capitalist networks and organisations working for food sovereignty (Kindling Trust, 2012; Bauwens, 2014).

This chapter discusses a selection of case studies that are operating in the third spaces of food production and consumption:

- Manchester Veg People (Manchester, UK): an MSC of growers, workers and buyers from restaurants, cafés and the University of Manchester who are growing, trading and educating about local organic food;
- OrganicLea (London, UK): a workers' cooperative growing food following permaculture methods and doing direct selling;
- Actyva (Extremadura, Spain): MSC of producers, consumers and a variety of workers, including media and PR professionals who support the promotion of good farming practices;
- Catasol (Aviles, Spain): an MSC that is also a workers' cooperative; all members share the ownership of the land;
- Central de Abastecimiento Catalana (Catalonian Supply Centre, or CAT): the food branch of the Cooperative Integral Catalana (Catalonia, Spain);
- Esnetik milk coop (Bilbao, Spain): MSC of sheep shepherds, workers, organisations, buying groups and individual consumers.

Out of the six, five are registered as MSCs and four of them were presented in detail in the previous chapter; OrganicLea is a workers' cooperative with an interesting model of production based on permaculture and high worker/stakeholder and political engagement. OrganicLea's workers' cooperative status is a rare legal form in UK farming (one of the only ones of its kind in the country at the time of writing). OrganicLea will also be referred to as an MSC, given its direct selling practices and close engagement with consumers.

A list of attributes that characterise third spaces can be found in Table 9.1. The next section compares MSCs' practices with the listed attributes in order to assess the extent to which they can be considered as promoters and recreators of third spaces.

Informal cooperative behaviours that increase connections between stakeholders with different identities

Informal cooperative practices are an important element of the culture of these MSCs; the conviviality and commensality of work, eating and cooperating are promoted and reflected in different ways: celebrations, open days and shared-labour days are frequent happenings to encourage regular interaction between workers, growers and consumers members; many also offer volunteering opportunities. Some, such as Actyva, have a strong communications strategy, providing a media platform for their producer farmers to connect with the outside world and learn to take pride in the value of their sustainable practices. The significance of the platform in supporting small producers and connecting them to consumers is reflected in the story of one of Actyva's producers with the last goatherd of his region, who works in a beautiful but very isolated spot and who has accrued a considerable social media following interested in his didactic tweets and photos of the valley where he herds his locally bred flock and makes cheese following traditional methods.

The informal cooperative behaviours were quoted by some participants as also being part of the rural culture they exist in, emerging from shared spaces; shared time – past, present and the expectation of a future together – modulates relationships and information sharing and decision-making processes (Ostrom, 1990; Polleta, 2002), as this quotes reflects:

> [S]ome are very used to meetings, and to discuss things and make decisions; the others aren't, the others are more like "let's meet, we have a barbecue and we can talk about it then" [. . .]. And I think it is because in reality villages do not have decision-making spaces as such, what they had were breaks from work that used to coincide with collective festivities and that was when deals were agreed [. . .] in villages there is another culture, in areas of a more rural character, because people see each other on a regular basis [. . .] it is very difficult to separate decision making spaces from leisure spaces and gossip spaces.
>
> (Actyva member)

These informal behaviours also take place across regional and national borders. Despite being in two countries, some of these cooperatives connect through wider events such as the gathering of MSC cooperatives in Spain, where Actyva and Coperativa Integral Catalan (CIC) coincided. In the UK, a connecting forum is the Oxford Real Farming Conference (ORFC), an increasingly popular and well-attended fringe event organised by the Campaign for Real Farming. Held in Oxford every January, the ORFC takes place at the same time as the more conventional Oxford Farming Conference where the big players of the food system meet (Oxford Real Farming Conference, 2016). At the alter-ego event, the ORFC, growers, activists and academics working on creating alternative food systems come together to network and share ideas, concerns and information, as one of the organisers told me:

> [I]t is quite a good example of what we mean by an alliance, because everyone is doing their own thing but they convene long enough to talk to each other and then they set up relationships between themselves of an ad-hoc type like Actyva and OrganicLea.
>
> (ORFC organiser)

These diverse expressions of informal cooperative behaviours result in deviant mainstreaming, as they follow Goodman and colleagues' proposition where they argue that AFNs based on relations and processes, rather than on fixed standards – such as the Fairtrade label and organic – are less likely to be co-opted and appropriated by the conventional food system (Goodman et al., 2011).

Governing third spaces and diverse memberships

MSCs have a diverse membership, with different stakeholders but also different types of membership option, with some cooperatives allowing members to volunteer hours instead of paying with money. Farmers are considered *prosumers*, because they contribute goods and services but also play a consumer role. Esnetik offers an interesting case, as its membership includes individual members and buying groups but also six legal entities, including a local council, an NGO and a rural development organisation. Initially, those organisational members were supposed to be there in an advisory role to offer support with the governance and business model, but they have ended up placing orders with the coop and thus becoming consumer members too. Some of Esnetik's shepherds regularly buy a lot of produce from the cooperative, reinforcing their dual role as producers and consumers (Ajates Gonzalez, 2017b).

In line with the theoretical features of third spaces, new terms and identities emerge in these cooperatives. An example of a new term is the case of the label neo-rurals, used to refer to urbanites moving to the countryside to start a new life in farming and, as such, embodying a new identity:

> [A]n ideologised neorural is one who has built around ecological agricultural or agroecology, not only an approach to production, but also their cosmovision, to interpret society, a former space perhaps filled by other ideologies, right? Those people are perhaps not as enthusiastic with

BBBF [Actyva's Project], but people who are more in the periphery of the agroecological world so to speak, that have not made of agroecology an identitary element, so yeah, those people love it, that there is information but also an informal atmosphere, combining quick and immediate advice to farmers who need it with a medium–long-term plan of transformation, not only of agriculture, but also of society.

(Actyva member)

Others, such as Esnetik, have emerged from the needs of shepherds who did not fit the logistic requirements of big industry players. As such, all producers in Esnetik were on the margins of the dominant milk sector: they were either not on the route of the tracks collecting milk – which meant they were offered low prices for their milk – or their production method was too traditional (e.g. milking by hand), not fitting the rewarded model of more litres at lower prices. These producers on the edge were able to be "rescued" by Esnetik, their same practices being relabelled, from being "disadvantaging" to becoming their unique selling point and asset in a cooperative with a different set of criteria from the ones they were previously assessed against.

When researching this diverse membership model (producers, workers and consumer members), the question of how MSCs can deal with this high governance complexity arises, as they have to take into account different stakeholders' views and objectives. A way to manage different opinions and bring consumers and buyers into the decision-making process is through weighted voting (e.g. MVP) or weighted representation on the board (e.g. Catasol or Esnetik), which helps these cooperatives to control power concentration in a single type of membership. According to their constitutions, all the case studies are set to have decision-making by voting; however, in practice, all of them aim for consensus as their preferred option.

Social movement: putting the movement back into the cooperative legal form

All of these MSCs follow multilevel cross-cutting collaborations at different levels, inspired by the "act local, think global" motto. They believe that social transition comes from social movements and the social base. They have horizontal and vertical links with other cooperatives, movements and civil society groups, both national (e.g. Campaign for Real Farming in the UK, GRAIN in Spain) and international, such as Via Campesina. In the UK, both case studies are active members of the Land Workers' Alliance, a relatively new trade union for small- and medium-scale farmers, challenging the current system that favours large monoculture farms. Actyva recently made attempts to develop a new initiative based on what could be termed a "local to local" principle, under the title of *Edible Organic Europe*, trying to set up partnership with other European cooperatives sharing their world-view and objectives in order to start a trade relationship within the solidarity economy they exist in and want to develop.

Esnetik in Bilbao is working with NGO Emaús to develop a Fairtrade label for its products that would be a first of its kind for growers in Europe. An interviewee from Via Campesina shared his concern about how a label could increase the absorption of Esnetik products in conventional retail chains and export markets. As Chatterton and Pickerill have pointed out, through these everyday challenges and contradictions, "activists are constantly border crossing between the familiar and unfamiliar, the world they are stuck in and cope with, the world they are against and resist, and the world they dream of and work towards" (Chatterton and Pickerill, 2010:487). Esnetik also organises *mercados eticos* (ethical markets); an Esnetik member told me what they consist of: "to organise a market and make a lot of noise to get people's attention, to explain the problems in the sector, try to raise awareness and change a bit consumption habits" (Esnetik worker).

This initiative links to other innovations that try to reduce co-optation by aiming to create new commons and markets embedded in negotiated normative frameworks shared by producers and consumers. Van der Ploeg has discussed how these new markets mostly emerge at the interstices – places where the functioning of large commodity markets is failing and not covering needs (van der Ploeg, 2013:84).

Other authors have warned how alternative initiatives can be labelled as too radical, often presented as practices far removed from the average consumer (Ajates Gonzalez, 2017b). In a similar way, Sullivan and colleagues argue, in their study of social movements, how any practice or organisation gaining significance in contesting and escaping the structuring enclosures of dominant regimes becomes labelled as "uncivil" (Sullivan et al., 2011), when, in fact, many would argue the "uncivil" are the defenders of the status quo. The case studies selected move from the civil to the "uncivil" and, in some cases, from the legal to the illegal. This is the case of the CIC, the overarching network of the Catalan food cooperative CAC, a well-organised network working to transform, not adapt to, the capitalist system.

The CIC was founded with money swindled from banks by activist Enric Duran (Kassam, 2014). From 2006 to 2008, Duran borrowed about half a million euros from banks, with no intention of paying the money back; instead, he gave the money away to fund and strengthen community-led initiatives on the margins of capitalism, because "capitalism won't allow us to create alternatives", and so the aim of the group he is part of is to create better parallel financial, health, food and education models, and so on, that "make capitalism become the marginal option" (Duran, 2016). In 2009, Duran was arrested and, after an anonymous supporter paid his bail, and facing eight years in jail, he became a fugitive and has been in hiding ever since. The CIC is working closely with the P2P Foundation to develop FairCoop, a global coalition of Open Cooperatives and a decentralised crypto-currency called FairCoin. This contrasts with the strict localised view of the CAC that only operates in Catalonia, balancing in a third space between the local and global action.

In all these cooperatives, people from very different backgrounds – many not from farming backgrounds – come together, creating diverse and unlikely groups with a wide range of skills and experience that would be difficult to replicate in more conventional food circles.

It is interesting to point out that some of the producers interviewed had been employed in workers' cooperatives not related to farming, before becoming growers; others had shared experiences in trade unions or the Green Party, bringing a new cooperative perspective and experience to agriculture that are very difficult to find in conventional ACs.

The diverse backgrounds of these members, some new to farming, encourage them to question and challenge stereotypes of how farmers look and sound. Perhaps for this reason, these cooperatives are becoming centres of convergence for different issues around land, crop varieties, breeds, seed varieties and control and gender (and gendered Spanish language). An example of their efforts to preserve local breeds comes from Esnetik's requirements for producers to use the local *lacha* breed of sheep; a similar approach is also taken by Actyva.

These spaces that foster flexible identities for members who can be at the same time producers and consumers illustrate the embodiment of the difference between what Bobel (2007) calls "doing activism" and "being activist"; many of these members are doing activism without self-identifying as activists, which increases the potential impact of these initiatives and engages a wider range of participants (Chatterton and Pickerill, 2010). The impact on attitudes of these wider interconnections was noted by some of the interviewees:

> [F]armers get more interested in the civic movement as they are exposed to the outside of their farms and their rural areas; those farmers want to have the more authentic governance system in the cooperative movement.
> (International Cooperative Agricultural
> Organisation representative)

These MSCs also aim for diversity in their offer to members of services and products beyond food items, especially the Spanish ones. The English cases focus more on offering workshops and facilitating new entries into farming, as is the case with the FarmStart programme run by the Kindling Trust, the organisation instrumental in the setting up of MVP. FarmStart courses are now being run to help other people set up similar projects in other regions. OrganicLea was the first one to follow the lead and now runs a sister FarmStart project in the Lea Valley in London. The CIC also coordinates a network for education and banking, among other projects. This more inclusive approach to cooperation that aims to cover different needs and dimensions of members' lives is more in line with the original vision of the early cooperators (Birchall, 1994; Münkner, 2004). Moving away from the specialisation trend that runs through conventional agriculture is another expression of deviant mainstreaming.

Consumers' evolving attitudes to food as well as buying and consumption routines are given space to change and evolve when they participate in these initiatives. These cooperatives are facilitating the passage from individual reflexivity to the collective action needed to resolve contradictions between values and patterns of daily life in conventional food systems that move the action from individuals to collectives (Brunori et al., 2012; Anderson et al., 2014; Gray, 2014b), facilitating a shift from assumptions to reflections about food. Bringing together different groups of stakeholders under the same umbrella for a common benefit threatens what Ulrich Beck observed as being the basis of the neo-liberal project: a conception of divided citizenship (Beck, 1992). In these MSCs, consumption is seen as a new political space where "the political possibilities of consumption (are) less than the revolutionary overthrow of capitalism but more than merely a niche marketing opportunity" (Goodman and DuPuis, 2002:18).

Some, such as Actyva and MVP, are trying to work with *less activist* consumers on a more regular basis, through their relationship with buyers in the case of MVP and developing a "social membership" for members at risk of inclusion in the case of Actyva. Others members shared their concerns about only reaching the most conscious consumers, and, aware of this point, all the Spanish case studies are expanding their offer of goods and services to present themselves as a more serious and wide-ranging alternative way to cover needs, starting with, but going beyond, food into other household products (e.g. Catasol selling environmentally friendly cleaners and beauty products).

When asked whether these networks complicate cooperative arrangements or instead help highlight the finite natural resources that are being depleted and the effects on the least powerful groups of society, one of the civil society organisation representatives made the following optimistic reflection:

> Well, I do not know, but the ability, we are saying that we live in a finite world, right? The infinite ability of our creativity, that is what gives an answer to everything, and that, we are not going to run out of. That is our secret, that there are infinite possibilities to reinvent the model.
>
> (Soberania Alimentaria Magazine rep.)

Legal form: multi-stakeholder models opening up new ways of economic reproduction

Cooperatives exist in capitalist economies and, in order to trade and protect their members, they have to incorporate as legal entities. Cooperatives are often presented as operating in a space in between the private and public sectors (Birchall, 1994). In this section, I explore how the case studies negotiate their legal existence with their vision for distributed power and democratic participation in economic activities.

Apart from OrganicLea, registered as a workers' cooperative, and Actyva, the rest of the case studies are registered as MSCs. Having a formalised legal structure can be seen in itself as a risk of co-optation and corporatisation (Mulqueen, 2012), but, at the same time, it can act as a protective tool that more informal food initiatives cannot make use of when conflict arises. As one academic interviewed put it:

> [G]overnance [. . .] and structures and constitutions (they) become very important at times of crises [. . .] when the informal cooperation [. . .] is threatened or breaks down, how do you settle those things? I see it as ice on a river, the ice kind of reflects the river and if the river goes away and you just have the ice, it then becomes very fragile, if you just had the structure, but without the substance that is supporting it, then it becomes fragile, open to demutualisation, to capture, to the kind of corruption we had in the coop bank, etc and in the coop.
>
> (UK academic)

Legally incorporating as cooperatives convert these groups of cooperators into trading bodies (Mulqueen, 2012) that exist in neo-liberal capitalist economies. They are very aware of that, and some have come out with mechanisms to create their own economic spaces, or, as one of the ORFC organisers put it: "in the short term farmers have to create economic micro-climates where they can start doing good things, protect themselves from the neoliberal nonsense of the global market".

An example of these microclimates can be found in Actyva's and CAC's acceptance and promotion of alternative currencies to pay for services and goods. The alternative currency community is another platform they work and link with. They also have inter-member trading and partnerships with buying groups as strategies that allow them to create new spaces to exchange their produce in more controlled conditions.

Actyva has adopted three interesting criteria to avoid attracting members with a sole interest in financial benefits: it is not-for-profit; it has capped salaries at 150 per cent of the equivalent in the normal labour market; and, third, it has a cooperative structure and a relational way of working, with democratic decision-making and a list of ethical and anti-sexism principles that would deter those looking for a quick and easy profit. In Catasol, all workers are on the same salary rate – difficult to replicate in more conventional models. However, as pointed out by Goodman et al. (2011), these principles complicate its financial sustainability and reproduction, as one of Actyva's member expressed: "it is like being on a tightrope between pragmatism and idealism". In order to survive financially, some of these cooperatives have started to sell their produce to non-members (e.g. Catasol, Esnetik and MVP). Actyva's producers sell part of their produce to other buyers outside the cooperative. Nevertheless, when selling to shops, they aim for like-minded enterprises selling products in line with their principles.

These cooperatives also have different ideas of growth compared with conventional ACs. Some prefer to limit their growth to a specific geographical area, such as the CAC. Others, if successful, see themselves growing horizontally into sister cooperatives, as is the case of MVP. All the cases value the flexibility of the MSC model (and the workers' cooperative model in the case of OrganicLea). A CIC member explained how they opted for the MSC legal form as it offered them a more flexible way of existing in the capitalist system while preparing a transition to overcome it.

An innovative move difficult for large retailers to replicate is Esnetik's introduction of double labelling, which consists in labelling all produce with a breakdown of the price and percentages that go to producers, packaging, processing and commercialisation (Ajates Gonzalez, 2017d). Members reported their pride in the supermarket-proof resistance offered by this strategy. Aware of how what they called the "agro-industrial model" has absorbed many of the initiatives and language of AFNs, such as organic and local, they were quite confident that supermarkets would not be able to incorporate double labelling, as it would reveal their own bad practices.

These findings support the counter cooperative-degeneration argument put forward by WIRC and back up the validity of the concepts of *deviant mainstreaming* and *incremental radicalism* in the context of AFNs (Arthur et al., 2008). All the legal and economic strategies discussed so far are employed by the case studies as mechanisms with several objectives: reduce co-optation, preserve their alterity while maintaining their economic reproduction, and spread their incremental radicalism. These objectives relate to Goodman and colleagues' concerns over the usual claims in the literature about how the transformative potential of AFN experiments "rest on the premise that the alterity of locality food networks is separable from the processes of their economic reproduction" (Goodman et al., 2011:78). The authors also point out that many AFNs are presented as "oppositional", even when they still rely on capitalist market relations and/or the state for their reproduction (Goodman et al., 2011). However, in the case of these emerging niche cooperatives, their deviant strategies and incremental radicalism are opening up small spaces of possibility in opposition to the dominant agri-food regime, to avoid "selling out to capitalist conformity" and find ways to provide "the economic security to perform and propagate these ethical values effectively" (Goodman et al., 2011:245).

In summary, the evidence suggests that the multi-stakeholder legal form is a means to operate in a capitalist economy, flexible and accommodating enough to be a reflection of these groups' founding principles, philosophy and objectives. The governance structure and practices associated with their multi-stakeholderism and worker democracy mean that the legal form is also being used, at the same time, as a tool to reduce co-option and create economic micro-climates.

Conclusion: distributed governance and economy as methods of resistance

This research has demonstrated that avoiding co-optation is an important objective of MSC in AFNs, which is sought to be realised with specific intentional strategies. The chapter has analysed the practices of six MSCs, based in Spain and the UK, that are attempting practices opposed to industrial food markets and to the mainstream model of agricultural production and coopera-tion. The case studies are part of a new wave of cooperatives born during the worst years of the 2007–8 financial crisis.

The concept of third spaces has proved useful in unveiling and unravelling their strategies towards co-optation avoidance, as well as unpacking the com-plex overlapping of issues and interconnections taking place in these MSCs that blur the fuzzy borders between producers, workers and consumers. As mem-bers, they all consume, and many of them contribute – in different ways – goods, labour and services to the cooperative. In these spaces, there is a more holistic representation of farmers as people with different hats and different needs, people who produce but also consume, challenging simplistic perceptions of growers. For farmers themselves, it is an opportunity to reflect on their own identity as consumers, on what they buy and how it affects their own farming practices.

Similar to processes of food policy stretching (Feindt and Flynn, 2009), these cooperatives are stretching both their mission (moving beyond a single-group membership and becoming actors of transformation) and the spaces they rep-resent: they exist as work spaces, learning spaces, but more noticeably, they are generating third spaces for cooperation where consumers, workers, buyers and producers can come together, rethink, produce and reproduce alternative ways of covering their needs. They are relational, open, internally diverse and externally stretched out. In some instances, they are stretching to uncomfortable spaces outside their comfort zone – for example, having to negotiate their rela-tionship with administrations and state subsidies.

Another strategy used by the case studies to resist co-optation is the adoption of more reflexive network governance approaches (Pirson and Turnbull, 2011; Turnbull, 2012 and 2013). The dispersion of power offered by distributed gov-ernance is only an internal innovation within the cooperative. However, these MSCs are also trying to be part of, and grow, a new economy moving beyond the private/public, centralised/decentralised dichotomies, towards a distributed model. They are recreating cooperatives as historic places for education, innova-tion, transformation and communication flows; while, on the one hand, they bet on a re-localised model of cooperation, on the other hand, they link their model to efforts directed to bringing about wider transformations beyond their regional borders. The most global initiative discussed is FairCoop, CIC's project to create a global coalition of Open Coops.

As cooperatives, they are reverting to the original "world-making" vision of the movement, away from the tame "shopkeeping" and "divi-dend" versions of it. They try to distance themselves from simplistic identi-tary "back-to-the-land" and "radical ruralities" labels, even though some of their members are "neo-rurals", which creates some tensions. They have a focus on progressive distributed economic models, based on creating decent employment opportunities and livelihoods in rural areas (especially key in the Spanish case studies). As one of the interviewees put it, they are examples of the infinite creativity of people trying to organise to meet their needs and reinvent their food systems.

The findings showed that a new wave of emerging niche cooperatives are opening up spaces of possibility in opposition to the dominant agri-food regime (Goodman et al., 2011), with three shared underlying objectives: to acknowledge and reduce the risk of co-optation, to preserve their alterity while maintaining their economic reproduction and to spread their incremental radicalism. Drawing on the concept of "deviant mainstreaming" and "incremental radicalism" (Arthur et al., 2008), evidence from the cases studies shows how that degeneration of ACs is not a contingent fate, but is not an easy one to circumvent. These MSCs are on a tightrope between adhering to their principles, surviving as enterprises and being able to advance and realise their visions of alternative food systems, the economy and society.

These ACs are organising both internally and externally, in line with more place-based and reflexive governance approaches that focus on processes and relations rather than on standards that are more likely to be co-opted, as has happened, to some extent, to the organic and fair trade movements. These initiatives are normalising less common (such as sharing farmland) and less conventional ways of working and cooperating. Through their activities, they are also trying to resist current processes of abstraction of the supply chain and agricultural investment, which are being increasingly disembodied in hedge funds and globalised multinationals with a highly untraceable subsidiary structure that obscures accountability. For this reason, it can be argued that these emerging cooperatives are opposite models to conventional farmer cooperatives that use cooperation as a means to perform better in the current system, without challenging it or attempting to transform it. It remains to be seen if, as one of the participants stated when asked about the future of these MSCs, "today's niche businesses are tomorrow's staple", or whether their efforts will continue to operate in in-between spaces.

10 Theoretical implications

A new integrated framework for deconstructing agricultural cooperatives

This chapter offers a grounded theoretical conceptualisation of cooperatives based on and interested in identifying and providing a framework for the analysis of patterns of action and interaction between actors and coalitions of actors and their role in shaping sustainable production practices.

Despite a wide range of diversity, ACs are developing an increasing market-oriented character aligned to the needs of large processors, retailers and globalised trade. Evidence suggests this is a Europe-wide trend (Bijman et al., 2012). This has had tangible effects on governance, the transfer of power to non-member managers and the prioritisation of financial over social objectives. The dominant strategies used by these large ACs to both compete in the European agri-food sector and benefit from EU food and agriculture policy include growing through mergers, acquisitions and internationalisation to concentrate power and be able to compete with private companies. This consolidation process is being supported and promoted by the governments of the two country cases covered: Spain and the UK. In Spain, this policy aim is transparent and clear, as the new law for the consolidation of the AC sector shows. As described in Chapter 7, the limit of withdrawable shares for UK cooperatives has recently been increased to £100,000, despite warnings this might benefit only the biggest and richest ACs' members and might endanger democratic practices if those members who are able to invest higher sums are given more power and influence. These trends show how these governance models can become very far removed from both the cooperative model and principles, and evolve to resemble and copy the strategies of IOFs. Another tension in the sector is the increasing growth in the number of POs across Europe, often formed with the sole aim of tapping into subsidies to cover the costs of pooling and marketing produce.

For Cogeca, the body representing ACs in Europe, sustainability is mainly seen in terms of effective production methods and profitability, as its support for sustainable intensification and GM feed suggests (see Chapter 5 for more details). The focus on exports, which many ACs across Europe are increasingly adhering to, is also contributing to intensive monocultures of export crops and food miles. Anecoop is a very clear example of this trend in Spain, growing water-thirsty crops in Almeria (the driest region in Europe, with the only true desert climate in the continent) and exporting nearly 90 per cent of its production to other countries.

The case studies presented in Chapters 8 and 9, on the other hand, suggest that, although the dominant model for ACs has clearly developed very permeable boundaries and is adopting characteristics and practices typical of POF models, there is still a lot of diversity and innovation taking place in agrarian cooperativism. A new generation of cooperatives are exploring innovative forms of governance and aligning themselves with solidarity economy principles and practices. Alongside the long-established smaller ACs and the more corporate ones, there is a growing number of producers who still see value in using the cooperative legal form as a tool to formalise their alternative ways of operating in the food system (Manchester Veg People, 2014; Ecological Land Cooperative, 2015). These new cooperatives are firmly embedded in AFNs and the relocalisation of food production in conjunction with the global food sovereignty movement.

Going back to the original theoretical framework, this research has provided an insight into the extent to which dominant ACs adhere to the Productionist Paradigm defined by Lang and Heasman (Lang and Heasman, 2015). Although most of the ACs presented as case studies are still composed of family businesses (e.g. Farmway, Anecoop, Esnetik, etc.), they are, however, embedded in the industrial and globalised model of food production to very different degrees. This finding is in line with van der Ploeg's distinction between capitalist and peasant farming (van der Ploeg, 2008). Many of the interviewees from MSCs defined themselves as peasants. Food sovereignty activists are reclaiming the word peasant, aiming to remove any negative connotations such as "unevolved" or "unprofessional" from this term. At the same time, advertently or inadvertently, they are adding new layers of political meaning to the word, as well as connotations relating to the agro-ecological farming methods adopted by these "new peasants". The variety of ACs studied reveals the tensions in the sector and the resistance to the increasing degree of assimilation by the "Empire", defined by van der Ploeg as the concentrated corporate control of the global food system and the dominance of profit-seeking and not sustainable practices (van der Ploeg, 2008).

As discussed in Chapter 4, the theoretical benefit of combining the frameworks of the food policy paradigms (Lang and Heasman, 2015) and van der Ploeg's new peasantries is the complementarity these theories offer to the different levels of analysis of this research: van der Ploeg's theory informed the organisational analytical level (the cooperative) to analyse how the different types of AC align themselves to or resist the "Empire". In conjunction, Lang and Heasman's food policy paradigms approach elevated the level analysis of ACs to the wider food system, providing a framework to assess their relationships with other actors (e.g. policymakers and retailers). The Open Coops framework proved a suitable lens for the analysis of MSCs. However, the findings and the fact that such diverse theories had to be brought together to be able to account for all the richness and contradictions in agrarian cooperativism reveal a theoretical gap and a need to further theorise the specific dynamics and role of ACs as key actors in the food system. How can such a wide range of realities and diverse experiences, from Anecoop to OrganicLea, get labelled under the same cooperative banner? The literature review, document analysis and interview data have revealed that dominant economic approaches to studying ACs are at their best limited, and, at their worse, damaging to the cooperative movement, farmers and the environment owing to their reductionist attempts to measure success. Informed by these findings and the noticeable theoretical gap that this research has highlighted, this chapter proposes a new framework based on a multilevel, multidimensional theorisation of agricultural cooperation beyond the dominant reductionist economic analysis. The chapter starts by discussing how the new framework relates to the theories that informed the research process (see Chapter 4) and tries to overcome different epistemological biases that emerged from the data presented: a reductionist understanding of cooperativism focusing on measurable financial indicators of success over social and environmental dimensions, coupled with a worrying trend to "export" this corporate model of ACs to developing countries. The new theoretical framework is broken down into two levels. The first level of the framework focuses on deconstructing the different elements that made up individual cooperatives. The higher level of the framework puts ACs in the context of the wider food system, analysing the relationships that any given AC might have with other actors in the food system (e.g. supermarkets, consumers, etc.). Finally, the concept of *cooperative sustainability*, which brings these two levels together, is introduced.

Overcoming biases in the study of agricultural cooperatives

Beyond quantophrenia: overcoming the reductionism of ACs

The literature review showed that the discipline with the largest body of academic literature on ACs is economics (see Gray, 2014a, for a detailed discussion on this point); there are numerous studies comparing ACs' performance with that of IOFs, mainly highlighting how ACs tend to do worse because they end up being undercapitalised and they have a demanding democratic governance structure. This approach reflects a methodological conservatism that has crept up on social science over the last 10 years (Denzin and Giardina, 2008), evident in governmental and funding agencies' preference for research that is quantitative, experimental and statistically generalisable (Tracy, 2010). This preference is the reflection of a wider obsession with financial survival in capitalist economies through the generation of higher revenues and strategies that secure competitiveness (Slaughter and Larry, 1997). The productionist paradigm that has been strongly embedded in global food policy since WWII reduced metrics of success of agricultural policies to pretty much a single and easily measurable indicator: yields (Lang and Heasman, 2015). Thus, the study of ACs is not the only food and agricultural realm in which the effect of quantophrenia is apparent, but is also palpable across policy and research approaches that create institutional lock-ins that reproduce reductionist and productionist paradigms (Vanloqueren and Baret, 2009).

Pitirim Sorokin, the first chairman of Harvard University's sociology department, coined the term quantophrenia in the early 1930s. The term refers to a fixation with, and a preference for, factors that can be easily measured, resulting in the misapplication of quantitative methods to sociology. The repercussions in public policy are obvious: potential policy solutions become reduced to a limited menu of options selected for their measurability (Sorokin, 1956).

Sorokin's criticism was not directed at quantification per se – he acknowledged the value of quantitative methods when applied to the right problems – but he argued that they should not be imposed by default and used as solutions looking for fitting problems. Paquet's concise explanation of Sorokin's argument deserves to be quoted here:

> The problem arises when the use of such tools becomes the basis of a *cult* roughly captured by the motto that if it cannot be measured, it does not exist. Such a cult distorts the appreciation we have of socio-economic phenomena, and this mental prison acts as blinders that have toxic unintended consequences for public policies when they are shaped by an apparatus thus constrained.
>
> (Paquet, 2009:2)

In the context of ACs, two decades ago, Mooney and colleagues identified how the reprivatisation discourse of neoclassical economics fuelled the depolitisation of US ACs in the 1980s and 1990s (Mooney et al., 1996). This coincided with the wider adoption of a universal framework for economic analysis. The choice of this framework also permeated academia and education on cooperatives. Kalmi has argued that this paradigm shift from institutional to neoclassical analysis resulted in "a neglect of the potential of cooperatives in addressing social problems" (Kalmi, 2007:625). By becoming the dominant discourse, neoclassical economics' theories become self-fulfilling through institutional design, social norms and language (Stofferahn, 2010).

Furthermore, the literature review presented in Chapter 3 shows how a reductionist economic approach continues to be dominant in the study of ACs (Sanchez Bajo and Roelants, 2011; Nilsson et al., 2012). Even New Institutional Economics perspectives that take into account human transaction dimensions still base their analysis on costs and have the research objective of quantifying the governance costs of ACs compared to IOFs (Iliopoulos, 2005). Management teams in ACs do not have instruments to measure or even estimate the loss of social capital that takes place when they undertake growth strategies. As that loss is not quantified in any way, it is not taken into account when decisions are made about the future of the cooperative (Nilsson et al., 2012). The cooperative principles have also been the subject of reductionist rewritings. Reviewing the history of the USDA's three AC principles, the USDA advisor, Bruce J. Reynolds, an economist, concluded the diluted principles were a "reduced form approach" developed through the "lens of economics" and prepared by economists to exclude values from the definition and identification of cooperatives (USDA, 2014:3).

Capital has completely subsumed the social and environmental spheres of life (Böhm and Land, 2009). However, these two spheres are especially inter-woven in food and farming practices: identity, place, animal welfare, commen-sality, biodiversity, abundance and so on are dimensions that cannot be reduced and that are complex to measure. Nevertheless, pervasive measurement and an evaluation culture are part of the way the capitalist regime reproduces itself, and food and ACs have not escaped its reach (Böhm and Land, 2009). In this sense, Allen has discussed the constriction of intentions that happens to sustain-ability discourses when principles become operationalised (Allen, 2004:18). The findings have shown a similar constriction in cooperatives' principles.

The complexity of integrating social justice at both the production and the consumption levels brings us back to the topic of structures and processes to measure sustainability using existing methodologies and understandings (see Chapter 3 for a detailed discussion). The inability to accurately measure social capital has been a weakness of proponents of cooperatives as a business model, as cooperatives always seem to lose out when compared with IOFs in organi-sational performance studies in the economics literature. Similarly, alternative food systems such as CSA, organic, agro-ecology and so on are frequently deemed to be expensive niches when, in fact, they are simply internalising many of the costs that, conventionally, food producers or retailers still do not include in their prices; in other words: cheap food is not cheap (Sustainable Food Trust, 2013). Therefore, the false sense of choice that the globalised and industrialised food system has brought to consumers is the deceptive shallow top layer of deep quicksands (Allen, 2004). The repercussions on ACs affect, not only the public and industry perceptions of the cooperative model, but also the definition of a cooperative and what a "successful cooperative" looks like, that is, the shaping criteria and indicators of success.

Western epistemology's three fundamental biases: ethnocentrism, androcentrism and anthropocentrism

Pérez Neira and Soler Montiel (2013) have put forward what they consider to be the three epistemological biases that the globalised food system suffers from, culprits of what the authors call a "Western epistemological crisis". Epistemology is a branch of philosophy that studies what is accepted to con-stitute valid knowledge and how it can be obtained (Vasilachis de Gialdino, 2009). The first of the biases proposed, the ethnocentric bias, refers to perceiv-ing and constructing understandings of other cultures and peoples as inferior, a trend that has been discussed in reference to some European ACs' impact in developing countries. Androcentrism highlights here the male-dominated, socio-economic, political and cultural structures of the food system. Finally, the authors highlight a third bias around anthropocentrism, reflected in the dominant approach of aiming to control rather than work with nature (Pérez Neira and Soler Montiel, 2013). These biases are intrinsically linked with the quantophrenic bias, as they define what is important and what should be repro-duced, in whose image. This research has discussed how these biases have permeated ACs in Europe, their relationship with their members and the version of ACs exported to other countries and cultures. Chapter 5 discussed how European ACs are exporting a top–down version of agricultural cooperativism to developing countries, with implications for small farmers, gender issues and sustainable practices. The proposed framework aims to overcome the above biases and offers a theoretical space and a language that can be of use to actors involved in alternative food systems and farming cooperatives when mapping, negotiating and understanding different cooperative endeavours.

An integrated theoretical framework for agricultural cooperatives

The cooperative triangle

Based on the findings from this research, this section puts forward a new framework that presents cooperatives as undertakings that can be deconstructed into four components that become expressions of cooperativism across four continua: legal form, governance model, social movement and informal cooperative behaviours (the last predate all other layers; Figure 10.1). These components are depicted in a cooperative triangle that has both theoretical and empirical value. First, it offers a theoretical representation of how certain elements associated with agricultural cooperativism can be easily absorbed into capitalist industrial food supply chains, while others offer more resistance. Second, it can be used to deconstruct and map any given AC and create a picture of where the organisation stands in each of the four dimensions.

Figure 10.1 Cooperative triangle illustrating the multidimensional character of agricultural cooperatives

Source: Author

In any given AC, each of the four sub-triangles will have a different size, depending on the cooperative's adherence to each continuum, namely: adherence of their governance model to cooperative principles, and their higher or lower degree of embeddedness in capitalist industrial food systems at one end or their alterity and commitment to the cooperative social movement at the other end. The other two continua refer to the implications for members of the legal form selected and the social capital among members reflected in informal cooperative behaviours, as opposed to other cases in which members do not know each other and use the AC as a service to buy inputs or market their produce.

The borders of each component triangle are seen as sliding scales on which different ACs' realities can be mapped, creating a cooperative triangle made up of constituent sub-triangles of different sizes, according to each cooperative reality. For example, a triangle mapping Farmway would have a well-defined legal sub-triangle with a smaller one for cooperative governance and less significant social movement and informal cooperative behaviours dimensions. In the case of a more informal buying group purchasing food directly from farmers, but without being legally incorporated as a cooperative, the sub-triangle for the legal form would be represented as a smaller portion of the whole triangle, whereas, in turn, the informal cooperative behaviours and social movement would occupy a more prominent place.

In the case of ACs, this multifaceted character is increasingly being fragmented by the mainstream food system, which is co-opting the less radical elements of cooperativism that can be easily absorbed without requiring a wider transformation of neo-liberal industrial practices. The next sections offer a brief description for each of the sub-triangles, as well as explaining how they relate to each other and the findings of this research.

Informal cooperative behaviours

This dimension refers to informal and unregulated expressions of cooperation among food producers. These informal cooperative behaviours pre-date all other elements of cooperatives that developed later; the benefits of cooperating in order to survive and grow enough food were recognised by farmers well before formalised cooperatives were created (Chloupkova et al., 2003). However, law and institutions often lag behind established social practices – in this case, cooperative practices in farming (Chloupkova et al., 2003; Sennett, 2012).

Informal cooperation in farming long preceded consumer cooperation. When consumer cooperatives became formalised in the mid nineteenth century, agricultural cooperatives followed, copying the model of the existing consumer ones. Fighting adulteration and achieving better prices were objectives shared by both consumer and farmer cooperators. In the last few decades, informal initiatives and associations trying to connect producers with consumers have mushroomed. These practices are only gradually being recognised and formalised into existing legal frameworks, for example, with the creation of the MSC model.

Legal

This component refers to whether an AC is incorporated or not as a legal body and, if so, in which form. The legal form and legal provision for cooperatives developed very differently across Europe, and still today, the legislation of cooperative societies varies enormously among countries (Bijman et al., 2012). Some countries have dozens of cooperative laws, such as Spain, whereas in others, such as the UK and Denmark, a specific cooperative legal form does not exist. This component triangle urges us to reflect on the legal form adopted by the organisation, if any, and the implications for members; for example, is the AC member-owned exclusively, or does it accept investments from non-producer members? What is the minimum investment required for new members, and how does this affect smaller farmers in the region?

For farmers considering different ways to cooperate, this dimension raises questions such as: What are the legal options available to farmers thinking of setting up a cooperative venture? Should they opt for the cooperative model that restricts their initial investment and probabilities of success, or should they opt for another legal form? Cogeca's representative revealed the tensions between POs and established ACs. In Spain, there are concerns about how the more liberalised producer associations might threaten the cooperative model if they become eligible to access the support that cooperatives enjoy at the regional level, without being expected to comply with the cooperative principles (Giagnocavo and Vargas-Vasserot, 2012). This reflects how the elements of ACs that enable the concentration of produce for processing or marketing purposes are favoured over the social elements related to the cooperative principles; new organisational forms of agricultural cooperativism ignore the latter to different degrees, such as the PO model that has been discussed extensively throughout the book.

This legal dimension also reflects the policy and political context. For example, when this research project took place, both Spain and the UK had Conservative governments (a Conservative/Liberal Democrat coalition in the UK, followed by a Conservative administration) promoting their understanding of the cooperative economy. Both countries underwent cooperative legislation changes at the time of the data collection stage. As discussed in earlier chapters, Spain introduced a new law in 2013 to promote mergers and acquisitions in the AC sector. In 2014, the Coalition government in the UK approved the Co-operative and Community Benefit Societies Act 2014, putting cooperatives on to the agenda and combining 17 different pieces of legislation that had not been updated for decades (UK Parliament, 2014). At the same time, many consumer and farmer associations that were not legally incorporated emerged. A new legal model, the multi-stakeholder cooperative, is bringing different types of member (consumers, workers and farmers) together under the same cooperative organisation.

Traditional informal networks of mutual help and assistance in farming communities (unincorporated cooperation) are in competition with legally incorporated forms of cooperation that allow access to subsidies. Some authors suggest that, by receiving funding from the government and the EU, ACs have essentially lost their independence from the state; that loss of independence has distorted and jeopardised the cooperative principles, which are increasingly disregarded in order for ACs to remain financially competitive and attract bigger farmers and more funds to survive in an extremely challenging sector (Zeuli and Cropp, 2004). Birchall has pointed out that capital needs autonomy from the state to care beyond profit (Birchall, 1994). When an AC is analysed, its approach to subsidies should be considered. Are members opposed to or in favour of seeking and receiving subsidies? On what basis? Subsidies were a bone of contention in some of the emerging cooperatives interviewed, creating dissensus among members with different opinions. For some small cooperatives, subsidies were perceived as the only way to get started; for others, as a threat to their independence.

Governance

Closely related to the legal form of a cooperative is its governance model; this component refers to the set of governance practices followed by any given AC, but also includes the effects of the wider food governance context in which ACs exist. In countries such as the UK where cooperatives can take many legal forms, their internal constitution is what defines cooperative governance and identity. This dimension also takes into account an AC's degree of adherence to the cooperative principles in the way it operates.

When an AC is analysed, the governance model should be closely examined. Does it follow the one-member-one-vote rule? Or is voting proportional to patronage? Are all workers offered the chance to become members and take part in decision-making processes, or is there a proportion of second-class workers? (For example, some ACs are outsourcing their operations to developing countries without allowing farmers overseas to join as members (see Berthelot, 2012)).

The case studies presented by this research have reflected a variety of arrangements, from those with one member, one vote, to newer forms of multi-stakeholder governance models where members still retain their vote, but each stakeholder group has a voting percentage, as in the case of Manchester Veg People and Esnetik (see Chapters 8 and 9 for a detailed analysis of these governance models).

Social movement

Challenging economic conditions are a fertile ground for cooperation; cooperatives, both at the consumer and producer levels, first emerged as a response to meet the specific needs of groups of people who struggled to access food or the land and inputs to produce it (Birchall, 1994). The first formal cooperatives that started in the UK and Ireland were not designed, or perceived by their originators, as an end in themselves, but as means to self-provide food and fund industrial cooperation projects in an attempt to transform society (Oakeshott, 1978). For them, the practical and financial gains from their cooperative transactions were not the final purpose; instead, they had a more ambitious vision that involved the creation of a new society based on cooperative principles. In contrast, the findings from this research suggest the current purpose of many ACs is to conform to, rather than transform, the capitalist industrial food system.

Can AFNs be considered a social movement, or are they something more modest (Allen, 2004; Kneafsey et al., 2008)? This is a question we can also ask about cooperatives. Cooperatives are part of an ongoing and evolving social cooperative movement; they are the expression and reflection of the problems and needs of the societies they exist in and they can only be as fair as the system they exist in. In that sense, cooperatives, as social movements, are dynamic rather than static; they try to resolve new challenges through bringing individuals together to cooperate based on their own understanding of what a cooperative is and what they should exist for (e.g. accessing subsidies or changing the world).

Next, the higher level of this new theoretical framework places ACs in the context of the wider food system, analysing the relationships that any given AC might have with other actors in the food system (e.g. supermarkets, consumers, etc.).

The double cooperative hourglass

Power structures and inequalities in the food system are often depicted by an hourglass (Figure 10.2) to reflect how a handful of powerful processors and retailers act as a bottleneck of profits and control between millions of producers and millions of consumers (Pimbert et al., 2001; Vorley, 2003; Patel, 2007; Thompson et al., 2007).

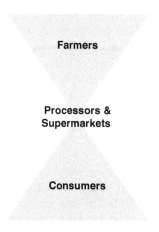

Figure 10.2 The food system hourglass
Source: Author

However, this hourglass diagram ignores the evidence this research has revealed regarding the powerful role ACs play in current food systems. Based on the findings of this study, a new, double hourglass is put forward as the second core component of the theoretical framework proposed. Building on the cooperative triangle, this higher level of the framework analyses ACs in the wider context of food systems.

Figure 10.3 focuses on illustrating power structures, not just for supermarkets and processors, but also ACs. It represents how many farmers sit between supply ACs at one end and marketing ACs at the other end. Obviously, some farmers are not members of any AC, but statistics show this is uncommon, and often farmers are members of more than one AC.

The double cooperative hourglass depicts how ACs can exercise power either in an outward direction, towards (a) the agro-inputs industry at one end or (b) the processors and retailers beyond the farm gate; or inwards, towards

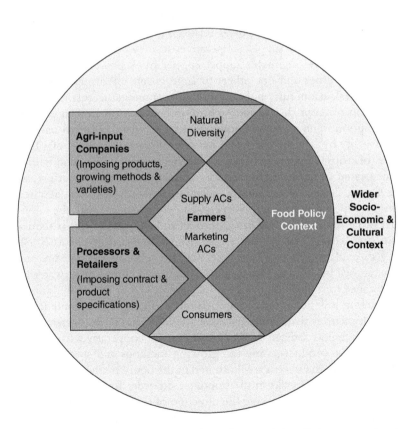

Figure 10.3 The double cooperative hourglass: are ACs offering outward resistance towards external powerful actors in the food system or inward pressure towards members?

Source: Author

their own farmer members. This framework proposes that, when one or more of the four components of cooperatives from the cooperative triangle are missing or become disjointed, cooperatives start to behave in a more corporate-like way and invert the direction of their power from outwards, as traditionally happened, to inwards. In contrast, some of the new cooperative models discussed in Chapters 8 and 9 are working hard to reverse the direction of farmers' collective power (and, in the case of MSCs, they do so in conjunction with consumers' collective power) to go outwards again, aiming to regain control over what they grow, how they grow it and how they sell their produce.

It is important to mention at this point that the food system is not linear, but circular (although not a neat, closed system, but a messy, complex one), and that there are waste and environmental impacts in every link of the chain, affecting the capacity of nature and biodiversity to reproduce themselves and, in turn, the food system. Therefore, this model does not advocate a simplistic linear vision of the food system that assumes infinite resources at one end and a limitless capacity to absorb waste at the other. This model is just zooming in and providing a closer look at the power relations in which ACs are involved.

The diagram shows how supply cooperatives sit between farmers and natural diversity. This level highlights supply cooperative's power to introduce certain varieties of crops and not others to large groups of farmer members. Numerous cultural, economic, political and environmental factors shape the narrow selection of crops that societies chose to cultivate for human or livestock consumption. The agri-input industry acts as a gatekeeper to nature's abundant diversity by commercialising a limited number of varieties of a limited number of crops (91 per cent of the 1.5 billion hectares of farmland worldwide are dedicated to mostly monocultures of wheat, rice, maize, cotton and soybeans; Altieri, 2007). Monoculture and homogeneous yields are favoured over multiple cropping and natural diversity. Closely associated with the intensive model of farming, agri-input companies introduce pesticides that reduce biodiversity and natural pest predators on farms (*Soberania Alimentaria*, 2013). The hourglass explains this trend by reflecting a reversal of the pressure and resistance exercised by ACs. In their origins, ACs exerted pressure outwards by offering resistance to supply companies selling overpriced or adulterated inputs or buyers offering low prices for produce (Birchall, 1994). The argument this research puts forward is that, nowadays, that resistance has been reversed, and external actors (retailers and agri-input companies) use cooperatives to squeeze farmers and impose products, varieties, growing methods and specifications. Power strongholds in the system are illustrated in the figure by the large players on the left creating bottlenecks in the hourglass. In order for it to serve as an analytical model, no arrows showing the direction of the "pressure" have been included, as its purpose is also to serve as a template to map actors and dynamics for specific ACs, each creating an individual hourglass with different numbers and size (importance) of players and power flows.

ACs deeply embedded in industrial food systems have developed governance structures and growth strategies that create unequal power relationships and lock-in situations for members. Additionally, their impact in developing countries and their increased co-optation by the dominant industrialised food regime (Friedmann, 2005), fuelled by practices of "co-opetition" (Walley and Custance, 2010; Pathak et al., 2014), show theoretical parallelisms with the conventionalisation thesis advanced by Goodman and colleagues in relation to the organic and fair trade movements (Goodman et al., 2011). Like the cooperative movement, both the organic and fair trade movements could also be considered social movements that can either have a legal stamp or not (certification in the case of organic and fair trade, or legal incorporation in the case of ACs) and can adhere to the original principles and visions of their movements or not. Many large supermarkets have created their own organic product lines that come from monocultures, with no intention to create more resilient and diverse agricultural systems – the original aim of organic farming – still following the same logic of bringing external artificial inputs on to the farm, just making sure that these meet the requirements for certification. The same conflict arises with the fair trade movement. Some authors have warned of how organic and fair trade labels have been appropriated and become mere selling points (Goodman et al. 2011; Griffiths, 2012). In this sense, the signifiers (e.g. the organic label or the cooperative name) become uncoupled from the signified (e.g. transformative organic principles or cooperatives) to mean something new (e.g. a set of standards in the case of organic or an organisation that facilitates the concentration of produce for easier processing, transporting and marketing). Fragmentation of what it means to be organic, fair-traded or cooperatively produced has given way for the most market-friendly dimensions of these (alternative) ways of production to be absorbed by the dominant food regime, becoming part of it and no longer aiming for a wider transformation, as was the root raison d'être of these original movements.

This institutionalisation of a standards-based, measurable approach to organic farming, fair trade or agricultural cooperativism is encouraged by policies that require groups of farmers to tick certain boxes to be eligible for certain certifications or subsidies. In this sense, there is a rupture from the original visions of transformation, as these movements are diluted by the substitution of their process-based approaches by standard-based ones of what could be called "allowable inputs" or, in the case of ACs, "allowable principles or practices". Current literature (both from academia and civil society) is discussing a similar danger of co-optation in the realm of agro-ecology (Levidow et al., 2014).

The role of policy

As reflected in the double cooperative hourglass, the wider food policy context shapes the relationships among actors and the power dynamics in which ACs are embedded. At the same time, the socio-economic context also shapes the way ACs embody the organisational dimensions illustrated by the cooperative triangle (Figure 10.1); for example, a national policy that facilitates the legal incorporation of cooperative enterprises will foster cooperative initiatives with a strong legal dimension. This section discusses the role food policy has in shaping the flows and pressures that take place in the double cooperative hourglass.

Agriculture has always been a sector characterised by heavy state intervention and a busy policy arena, where both private and public actors have been closely linked (Davey et al., 1976; Bijman et al., 2010); these links have created tensions around policy objectives that are often presented as conflicting, such as sustainability and high yields. Beyond the CAP, the current discourse in most European countries is based on the perception of the state as a facilitator and the need to encourage farmers to become more entrepreneurial and business-like. Lockie (2009) has pointed out that this strategy is a means by which government can transfer food production responsibility from the state to farmers, also transferring accountability.

With regard to ACs, farmers' decision to legally incorporate or be part of informal models of agricultural cooperation is shaped, among other factors, by the policy context, which can either hinder or encourage specific models of farmer collaboration. Owing to the complicated policy arena where domestic and international levels of food governance often clash (Lang et al., 2009), farmers find themselves having to take into account CAP subsidies, competition law regulations and EU directives. The consistent contradictory message is that farmers should both compete and cooperate, now promoted and encapsulated in the term "co-opetition" (Galdeano-Gómez et al., 2015).

According to the ILO guidelines for cooperative legislation, state intervention in cooperative affairs should be restricted to four functions: legislation, registration, dissolution/liquidation and monitoring of the application of the law by the cooperatives (ILO, 2005, in Henrÿ, 2012). The ILO guidelines take the premise that:

> [T]he main objective of a cooperative law [should] be to guarantee minimum government involvement, maximum deregulation, maximum democratic participation and minimum government spending by translating the cooperative principles into a legally binding framework for the organisation of self-determined self-help.
>
> (ILO, 2005:vi, in Henrÿ, 2012)

However, as has been discussed in Chapter 5, although on the one hand the EU is not able to dictate the legal form of farmer enterprises, on the other hand, through subsidy signals, it is actually fuelling forms of cooperation that neglect the cooperative principles, either by focusing on capital, not members, or by providing funds to POs and other looser cooperative forms. In this sense, EU policies have cognitive effects on the strategies of societal actors and their incentives to mobilise and/or build coalitions and collaborate (Skogstad, 1998). Those effects also take place at the national level, through domestic policies and legislation (Cairney, 2012).

Richard Fletcher (in Coates and Benn, 1976:179) pointed out that Britain "has turned to cooperation as a form of self-help at the times of economic crises". Change triggered by crises is a typical phenomenon in food policy. In the UK, for example, this is clear from the history of government-led food policies, especially during wars and farming crises such as BSE or foot and mouth disease (Lang et al., 2009). Sadly, as this research has revealed, paradigms such as productionism have an enduring impact on policy once they become institutionally embedded, and it takes a severe crisis or shock for proposals for change to be considered. This paradigm has not only created path dependency in food and farming policy (as discussed in Chapter 5), but has become the interpretive framework through which actors in key political institutions and organisations view the world and frame policy problems and potential policy solutions (Skogstad, 1998). In the case of cooperativism, ACs are no longer perceived by most as transformative projects, but as functional enterprises aligned with the aims and logistics of the productionist paradigm. However, as Hall, Skogstad, and Lang and colleagues have pointed out, crises jeopardise ideologies and overthrow paradigms (Hall, 1993; Skogstad, 1998; Lang et al., 2009).

Is there an ideal policy context that can foster a type of cooperative that can help farmers balance short-term and long-term objectives and gains and, in turn, inform a less short-term food policy approach? In fact, when ACs are considered as part of food and farming policy, their potential as helpful institutional actors that can help link the international, national and subnational spheres (an inclusive term for the less clear regional and local levels) becomes apparent. However, some authors have argued that many food policy issues are still at an early embryonic stage (Lang et al., 2009). Problem definition of food policy agendas may lead to agreement, but agreeing on the solutions needed to address those problems is not as easy. There is general agreement on the need to increase collaboration among farmers (problem formation), but less on the policy solutions. For example, agriculture policymakers and the food industry may agree on the problem of lack of collaboration but frame the solution within the neo-liberal paradigm – for example, with the creation of liberalised POs – as opposed to proposing the wider transformation of the food production system and broader economic relations. Depending on the perspective framing the problem, cooperatives can be put forward as a means to a wider transformative end or as a mechanism serving the current dominant system.

The notions of collaboration and cooperation are used as discursive concepts to promote concentration of produce and agricultural policy entitlements and subsidies, while still encouraging post-productivist agricultural policy measures. This is possible owing to the flexibility of the cooperative model and its scope for political accommodation. As the cooperative triangle showed (Figure 10.1), an AC can be legally registered as a cooperative but have nothing to do with the wider cooperative movement or, indeed, any food movements. However, this research has revealed that empty discourses of cooperation in farming are highly problematic: on the one hand, those discourses project a positive image of ACs and serve as a strength in lobbying for more funding for the AC sector. But that same image of the halo effect makes it difficult to identify at first sight the real impact of and underlying reasons for encouraging collaboration and cooperation in the food system on a big scale. Whereas this empty discourse can be useful at that level, it becomes a liability when it comes to practical implementation settings. This challenge around problem and solution framing is also common in multifunctionality discourses in farming, which highlights how the productionist paradigm is the underlying common denominator:

> For although multi-functional discourses go some way towards explaining the need for a transition from productivist to post-productivist agriculture, they do not set out how this process should be enacted, how organisational responsibilities should be assigned, or, crucially, specify implementation goals or endpoints.
>
> (Clark, 2006:347)

In a similar way to multifunctional and sustainability discourses, cooperative discourses do not explain how to reach the desired end: should we achieve sustainable food systems through GM and sustainable intensification? Or by promoting agro-ecology and food sovereignty? In the case of cooperation, should we foster cooperation among farmers through the ICA model of cooperatives? Or through vertical integration through contract farming between producers and large retailers? The research findings suggest cooperatives are just a vehicle to reach many diverse objectives.

What is the potential of the social model of cooperatives to help shape consumer demands (e.g. through shortened food channels, food hubs, etc.), as opposed to the dominant AC model, often based on long supply chains and foreign trade? Can or should policymakers tap into ACs and foster, rather than inhibit, their social aspects for wider gains? Can the dominant AC model help achieve both food security and sustainability goals and bridge issues of food safety, food control and food democracy? Evidence seems to suggest not. A high degree of polarisation in the AC sector, fuelled by a combination of subsidy policies and the concentration of power down the food chain (in processing and retail), is leaving behind small and medium ACs in favour of more financially competitive and larger ACs (Gray and Stevenson, 2008). In Spain, the government has started an overt battle to eliminate uncompetitive medium-sized cooperatives from the market in order to improve the efficiency of the Spanish agricultural sector (and the efficient administration of subsidy payments). At the opposite end, large ACs are benefiting from subsidies and rewards when they increase their size. Cartwright and Swain have explained how big players in farming are enjoying subsidies they do not really need or deserve by using a similar, nearly empty imaginary:

> The farm lobbies of the western European countries successfully mobilise political support for agricultural policies which disproportionately favour the big, "Northern" farmer by invoking the needs of the small farmer, the "peasant", for whom farming is becoming in reality increasingly marginal. [. . .] First, "North" and "South" are not accurate geographical terms. The "North" and the "South" are present in every country. In this sense, every

European agriculture has a dual structure, but the European Union has singularly failed to recognise this duality in its policy formulations.

(Cartwright and Swain, 2003:12)

Governments and civil society can encourage cooperation, but, at the end of the day, it is farmers who are risking their capital and way of living when joining an AC. As the academic from ISEC mentioned, many farmers feel trapped in ACs that have a model of intensive agriculture for exports, which represents a social and political issue, not a technical one. Technical and yield-related challenges were the main policy problem post-war food policy had to tackle, and it did so by focusing on edaphology, agronomy, biology, logistics, and so on, to increase production, reduce food costs and improve distribution. The current challenges facing European food policy are fundamentally social and political: how to change consumer habits to reduce waste, how to improve the health of the environment and the population, and how fairness and food democracy can be achieved for producers, workers and consumers (Lang et al., 2009). However, the productionist paradigm is still strongly rooted in and embedded in the current food-policy-making logic (Lang and Heasman, 2015), and the AC sector has not been spared its influence (Giagnocavo and Vargas-Vasserot, 2012). As depicted in the double cooperative hourglass (Figure 10.3), ACs can be used or can act as mere convenient shortcuts to concentrate produce and tap into large groups of farmers in order to: (1) sell them inputs more easily (useful for agri-industries); (2) buy their products (useful for large retailers); and (3) process subsidies or implement regulations (useful for policymakers and civil servants). In this sense, ACs' practices and new models of cooperation are not a straightforward reflection of the cooperative discourse itself, but also – or mainly, in the case of agriculture – the policy and institutional frameworks in which they exist.

Cooperative sustainability

This section introduces the notion of "cooperative sustainability" (CS) to refer to the sustainability of the cooperative itself, its essence, especially with regard to the social movement dimension of ACs, but also with interwoven environmental connotations.

With economic edge being the main *raison d'être* for a majority of ACs, cooperation becomes a specific means to an end, seeking continued short-term profitability, as opposed to cooperation as part of a social movement, with the final ambition of transforming society into a cooperative one (Oakeshott, 1978; Mooney and Majka, 1995). Large ACs can be made to fit two purposes: achieve more measurable and financial objectives as part of a process of professionalisation of farming cooperatives; and, second, fit the state-supported agricultural paradigm, by facilitating eligibility assessments for the allocation of farming subsidies. The double hourglass showed in Figure 10.3 helps illustrate this point. At one end of the food system, we have nature, with outstanding biodiversity. Supply ACs become the first point of contact between the consolidated agri-input industry and farmers. A handful of agri-input corporations control most of the market and, thus, control the varieties that are commercialised.

As ACs grow more focused on monocultures, they become more dependent on these companies and contribute to loss of diversity and increased "bio-terrorist threats" (Mooney, 2004:96). In this level of the hourglass, power imbalances already appear as, despite the high number of cooperatives existing in the EU, a very small group account for most of the trade, which is also the case for marketing cooperatives (Bijman et al., 2012). Large marketing cooperatives fit well within the logistics model of large processors and supermarkets that find it much easier to deal with one cooperative salesperson than 5,000 individual farmer members. In the context of sustainability, Copa-Cogeca has highlighted how the fact that the market does not remunerate the provision of public goods is an issue, and that financial incentives within and beyond the CAP are needed, as the CAP is neither a research, nor a climate change policy (Copa-Cogeca, 2011). In the meantime, Copa-Cogeca is adopting a more reductionist conception of sustainability based on sustainable intensification (Copa-Cogeca, 2011; Berthelot, 2012; COCERAL, 2015a, 2015b).

Thus, sustainability is not a result or an area of internal AC policies. Rather, different understandings of sustainability inspire different groups of producers (sometimes experienced growers, sometimes neo-rurals) to start new cooperatives. The conventional ACs use cooperation as a means to maximise profits and reduce costs. MSCs see cooperation as an end, as the right thing to do, not just with other farmers but also with consumers. As discussed in Chapters 8 and 9, a renewed focus on alternative approaches to cooperation and integration in the food system has given way to an innovative legal model: multi-stakeholder cooperatives.

In this sense, these new MSCs demonstrate a post-productionist approach, whereas the conventional ACs studied in Chapter 7 are still stuck in the productionist paradigm, with farmer members suffering and reproducing the problems caused by this model. Although MSCs are moving away from productionist definitions of success, sustainability and efficiency, they do not exist in a vacuum, and so tensions and compromises are ongoing.

The notion of CS brings together both the cooperative triangle and hourglass, as the data revealed that the factors weakening CS are bifold and inter-linked with issues of specific understandings of "professionalisation of farming" on the one hand and lack of cooperative education on the other, as discussed previously. Based on these data, I define CS as the ability or capacity for an AC to maintain the four dimensions indicated in the cooperative triangle, while also fostering environmental sustainability.

Former debates around CS have been mainly on the legal dimension proposed by the framework presented in this book, primarily debates around conversion to IOFs (Mooney et al., 1996; Dunn, 1988). The concept of CS contributes to this literature both theoretically and methodologically. Theoretically, the CS concept expands the debate to other dimensions of ACs, beyond their legal form, governance and investor profile. Methodologically, CS is a construct that can be evaluated, explored and further developed through the application of the theoretical framework proposed in this chapter as a tool for assessing individual case studies; the application can be done on either specific cooperatives or to assess the potential impact of policies relating to ACs and agriculture, by analysing how a particular policy intervention can impact CS based on the effects/implications for each of the cooperative dimensions proposed.

This research puts forward the concept of CS to suggest that, unless ACs succeed in integrating the four elements of the cooperative triangle (Figure 10.1), their model will eventually become something else, no longer cooperative and no longer transformative. The findings from this research suggest the key factors that could help achieve cooperative sustainability with wider benefits for the wider food system are:

1   redirecting pressure outwards towards large players in the food system, instead of inwards towards members;
2   overcoming Western epistemological biases of anthropocentrism, ethnocentrism and androcentrism; and
3   overcoming quantophrenic definitions of success.

Conclusion

This chapter has put forward a new analytical framework that emerged from a multi-method approach based on the examination of multidisciplinary literature, policy documents and diverse sources of new data. From the wide variety of data and informants consulted, a conclusion emerges: cooperation in farming can be on a sliding scale of (a) formality (from informal, "neighbourly" behaviours to legal incorporation that varies greatly across countries) and (b) adherence to the cooperative principles and the wider cooperative social movement, with multiple hybrid combinations in between.

Over the years, as a result of the above factors, components of ACs have become disjointed. The pressure that ACs were originally set up to exert outwards has become increasingly inverted towards cooperative members themselves, blunting their transformative power. Previously, ACs used to exert resistance against powerful actors upstream and downstream in the food system, by both protecting and empowering farmers to deal with the agro-input industry and with large buyers from processing and retail firms. This research has shown how certain elements of ACs can be more susceptible to co-optation than others; evidence presented showed how powerful actors in the agro-input, processing and retail industries can turn the advantage of having many producers grouped in a single AC to their own benefit. Agro-industries use ACs as an entry to point to introduce product lines and varieties; processors and retailers also benefit from ACs by imposing quality standards and logistics demands on a large number of ACs' members in one go, rather than having to negotiate with many farmers individually. ACs fit well the model of purchasing and logistics of the industrial and globalised food system.

The growth of larger, more corporate ACs reduces farmers' options at several levels: there are fewer ACs farmers can join, and, once they become members, they experience more top–down control from their ACs in terms of what to grow and how to grow it and face lock-in situations that are hard to escape. This control over farmers is even more critical for poor and illiterate AC members in developing countries, where neo-colonialism and the globalisation of food production are being crystallised in the creation of top–down ACs.

Research into this particular topic of co-optation of the cooperative model raises a number of difficult issues around ACs and highlights some uncomfortable – but potentially analytically fruitful – tensions between cooperative governance and cooperatives' survival. At the same time, the new theoretical framework presented provides explanatory power to differentiate between cooperatives as organisations and cooperative practices. The double cooperative hourglass (Figure 10.3) can help illustrate these ongoing tensions between the cooperative economic model and the market economy in which cooperatives exist. The critical theory put forward as part of this research also allows trends in labour relations and dynamics of production and power in European ACs to be analysed, as well as their impact overseas, a key issue in an increasingly globalised food system.

The proposed framework builds on and is complementary to Goodman and colleagues' conventionalisation thesis of the organic movement (Goodman et al., 2011). Other food movements, such as organic and fair trade, are also struggling with processes of co-optation. The framework reflects how ACs can be situated on a continuum of alterity, from those that are highly embedded in the dominant food regime to those that aim to create a new economic system, starting with elements of food provision. Therefore, ACs can be considered either part of the alternative food systems or embedded in the industrial food regime, depending on where on that continuum of alterity they sit, with many hybrid forms emerging.

The preference for vertical integration in the market economy and globalised food system has overtaken the horizontality and equality principles that formed the original roots of cooperatives. This chapter has put forward the concept of cooperative sustainability to suggest that, unless ACs succeed in integrating the four elements of the cooperative pyramid, their model will gradually become something else, no longer cooperative and no longer transformative. Strategies to achieve sustainability include redirecting pressure outwards towards large players in the food system, instead of inwards towards members, overcoming Western epistemological biases and quantophrenic definitions of success. For this reason, when promoting cooperatives as a way to achieve more sustainable and fairer food systems, we need to be careful and more specific about the types of cooperative and cooperation we want to promote.

11 Conclusions

Implications for agricultural cooperatives, food policy and alternative food initiatives

This closing chapter includes final conclusions, reflections and recommendations for further research. It starts by discussing the implications the findings have for the agricultural cooperative sector and wider alternative food initiatives. The chapter ends with a discussion of implications for farming and cooperative policy and reviewing opportunities for further research.

Implications for the agricultural cooperative sector

This book has portrayed twenty-first-century agricultural cooperation in the context of EU food and farming policy. By combining a thorough multi-disciplinary literature review of agricultural cooperatives and food sustain-ability with case study methodology, the book has covered a wide range of theoretical and empirical ground. Empirically, this book has documented the first analysis of food MSCs in Europe. The data and analysis provided have stretched binary understandings of the conventional versus alternative dis-courses dominant in food studies. Additionally, the book has attempted to unpack and theorise the diversity and possibilities of different cooperative forms in agro-food systems and how they evolve to adapt or resist wider pol-icy and power dynamics in the food system. As a result of its theoretical and empirical contributions, this study has opened up a new research agenda in the field of interdisciplinary food studies that will be discussed in more detail at the end of this chapter.

This research has aimed at providing a comprehensive history of ACs in Europe, analysing the challenges and contradictions faced by contemporary ACs. In the country cases, in relation to the historical dimension covered when tracing back the historical roots of ACs, the more popular element of the UK cooperative movement focused on food retail, whereas the Spanish tradition has always been more land-based; this difference makes social and cultural sense when considering where the main bulk of the labour population lived in each country (urban or rural areas) at the time the cooperative movement devel-oped, and what needs cooperative initiatives were able to cover (production- or consumption-related). Nevertheless, although there are still wide-ranging differences between Spanish and British ACs, owing to entrenched historical and socio-economic factors that are representative of the contrast between southern European and northern European countries discussed at the beginning of the book, this research has highlighted two trends common to both countries, and more widely, to European ACs: increased liberalisation and consolidation in the AC sector. This book has presented evidence of the general lack of engagement the dominant AC sector has with the wider cooperative movement. The theoretical starting assumption of this research was that ACs were once expressions of alternative food systems at the farming level, to empower food producers and tackle power imbalances in the food system. With industrial agriculture and globalisation, that alterity is being lost.

The wide variety of data and informants used in this research has revealed how cooperation in farming can come on a sliding scale of:

- formality (from informal "neighbourly" behaviours to incorporation into legal frameworks that vary greatly across countries); and
- adherence to the cooperative principles and the wider cooperative social movement.

The rich variety of ACs in the EU means there are many hybrids and multiple combinations in between. The findings from this research show that factors increasing the corporatisation of ACs in the EU are multilevel:

- cooperative level: lack of continuous education of members on the history, meaning and value of their cooperative and the wider cooperative movement;
- national level: policies aiming to increase the profitability of farming as an economic sector, including clear support for consolidation of the sector and larger farms;
- EU level: the CAP pushing for the PO model to concentrate demand and facilitate trade within the Common Market and beyond;
- global level: the introduction of agriculture into the World Trade Organization in 1995 and increasing demand from emerging economies are also pushing ACs both to become bigger players to be able to compete in their national markets and to export to foreign markets.

Certain elements of ACs can be more susceptible to co-optation than others. Evidence presented showed how powerful actors in the agro-input, processing and retail industries can turn the advantage of having many producers grouped in a single AC to their own benefit, using ACs as an entry point to introduce product lines and varieties (useful for agro-industries) as well as quality standards and logistic demands (useful for processors and retailers) to large numbers of farmers in one go.

This book has also discussed the role and potential of emerging MSCs to decentralise power and control over supply channels in the food system at a time when capitalism is concentrating, not only capital, but also its workforce (mostly net consumers of food) in cities. Data presented showed how the MSC vision is based on and requires politics of collective responsibility for it to succeed, as it aims to adopt more relational ways of trading, based on geographical and temporal connections. The findings from the case studies indicated that the MSC model is demanding and messy, reflecting a genuine and passionate attempt to realise all the criteria of multidimensional sustainability.

The MSCs presented in the case studies are aspiring to transform the food system following two strategies: first, by raising awareness of and reducing unaccounted externalities of food production. Second, by aiming for multidimensional sustainability, creating a multi-campaign space where efforts to address the new fundamentals are coming together (Lang, 2010). In contrast to mid-nineteenth-century class-based cooperativism, MSCs exemplify a much more diverse and diffused notion of cooperation, the multiplicity of their objectives being a reflection of the increasing policy stretching that global food governance is experiencing.

At the organisational level, if principles, governance and outward networks are cultivated, MSCs can become a connecting link between bottom–up initiatives and top–down food policies, a link with the potential to be scaled up by collaborative public procurement strategies (e.g. scaling up current projects around supplying school and university canteens). Their heterogeneous memberships and networks make them complex, but also give them resilience, contacts and a voice on different platforms of the multilevel governance arena of global food policy. The emergence of MSCs highlights the need for a post-productivist model of cooperation, one that acknowledges the challenges that consumers and producers are facing in the twenty-first-century food system, opposing the outdated approach to food policy that confines them to isolated silos and striving for integration, an essential condition to achieve truly sustainable food systems.

This research has identified theoretical parallelisms between the agricultural cooperative movement and the organic, fair trade and agro-ecology movements, which are also struggling between conforming to and transforming the dominant industrial food regime (Levidow et al., 2014). The new proposed theoretical framework situates ACs in a continuum of alterity, from those that are highly embedded in industrial and globalised food systems to those that aim to create a new economic system, starting with elements of food provision. Therefore, ACs can be considered either part of alternative food systems or embedded in the industrial food regime, depending on where on that continuum of alterity they sit.

The theoretical framework presented in Chapter 10 has a multilevel character to reflect the complexity of the food policy context ACs exist in. The first level of the framework, the cooperative triangle (Figure 10.1), focuses on deconstructing the different elements that made up individual cooperatives, namely, informal cooperative behaviours (that predate all other layers), legal dimension, governance model and level of connection with the wider cooperative movement. The second level of the framework, the double cooperative hourglass (Figure 10.3), puts ACs in the context of the wider food system, analysing the relationships that any given AC might have with other actors in the food system (e.g. supermarkets, consumers, etc.). Finally, the concept of cooperative sustainability brings the previous two levels together by referring to the sustainability of the cooperative itself, its essence and principles, especially with regard to the social movement dimension of ACs, but also with interwoven environmental connotations. The multilevel framework suggests that, unless the four elements of the cooperative triangle (Figure 10.1) are balanced, ACs will eventually become something else, no longer cooperative and no longer transformative, putting pressure on their own members rather than the powerful players they were originally set up to resist.

This new framework offers a new perspective to examine ACs and contributes an analytical tool that can be used for the evaluation of specific ACs or individual policies; more importantly, it introduces a new language to describe and analyse cooperatives in food and farming that those involved in alternative food initiatives can use to situate, negotiate and understand different cooperative endeavours. Can cooperatives resolve the dilemmas of living in an unsustainable food system? The evidence suggests that is not necessarily the case, but cooperation grounds the process, it adds coherence and injects a human perspective to the debate.

Theoretical, disciplinary, methodological, empirical and social implications

This section describes and reflects on the multilevel theoretical, disciplinary, methodological, empirical and social implications of this research. Delving into the topic of co-optation raises a number of difficult issues about the ACs as it highlights some uncomfortable – but potentially analytically fruitful – tensions between cooperative governance and cooperatives' survival.

On the one hand, it is challenging to be critical of cooperative organisations that, even if big and embedded in long industrial supply chains, are nevertheless allowing individual farmer members to continue making a living – even if often a meagre one – in farming. On the other hand, a solidary-critical approach (Favaro, 2017) is much needed in order to avoid idealising cooperatives that no longer adhere to their principles and ethics, to call them out and to encourage them to reflect rather than rest on their historical laurels.

In theoretical terms, this research has offered an alternative perspective to the economics discipline that dominates the study of ACs, presenting instead a mainly qualitative study from a food policy and rural sociology perspective. Additionally, a wide range of literature from a variety of disciplines was reviewed, converging complementary perspectives in the study of cooperatives. The findings of this research have shown how no one discipline, but rather a multidisciplinary approach, is required to start unravelling the complexity of cooperative enterprises and all the aspects they entail.

At the same time, the new theoretical framework presented in Chapter 10 provides explanatory power to differentiate between cooperatives as organisations and cooperative practices. The double cooperative hourglass model can help illustrate these ongoing tensions between the cooperative economic model and the market economy cooperatives exist in. The critical lens put forward also allows analysis of trends in labour relations and dynamics of production and power in European ACs, as well as their impact overseas, a key issue in an increasingly globalised food system. This framework builds on Patrick Mooney's work on tensions in ACs (Mooney et al., 1996; Mooney, 2004) and contributes to what Mooney called "a sociology of cooperation" (Mooney, 2004:96), aiming to resume the debate on the de/repoliticisation of agricultural cooperatives (Mooney et al., 1996; Mooney, 2004).

Since the creation of the European Common Market, ACs have been gaining greater institutional legitimacy and subsidies, advancing the cooperative model and presenting it as a business case claim rather than a sociopolitical agenda of transformation. For this reason, the conventionalisation thesis was also used to explain the political skimming of ACs. The double cooperative hourglass offers sufficient explanatory potential to be applied to the case of organics and fair trade, as those words were once mainly used to refer to movements and nowadays they are mainly applied to labels and certification. In a similar process of co-optation to ACs, the organic and fair trade movements did once empower farmers, but existing evidence suggests the certifications are now being used as a checklist of allowable inputs (in the case of organic) or allowable working conditions and price paid per product (in the case of fair trade); the labels once used to identify and represent the movements are, in many cases, becoming mere selling points for new "ethical" product lines for aware and affluent consumers (Goodman et al., 2011).

The new concept of CS put forward contributes to the literature on ACs and alternative food systems both theoretically and methodologically. Theoretically, CS expands the debate to other dimensions of ACs, beyond their legal form, governance and investor profile. Methodologically, CS is a construct that can be evaluated, explored and further developed through the application of the theoretical framework proposed in Chapter 10 as a tool for assessing individual case studies.

A potential future use of the cooperative triangle and double hourglass framework can involve its application to individual cooperatives. Additionally, the framework also proposes widening the range of perspectives used to examine ACs. It proposes a tool for mapping ACs' practices and to create a language that those involved in alternative food systems and cooperatives can use to negotiate and understand different cooperative endeavours. The four proposed dimensions can be seen as sliding scales on which ACs' realities can be mapped. Furthermore, the original application of the anthropological concept of "third space" has offered a new and original approach to study the alterity and resistance of alternative food networks.

Additionally, interwoven with the disciplinary contribution, the methodology offered a qualitative study of a subject, ACs, that has previously been examined mainly quantitatively and through a business performance lens.

The multilevel analysis of comparison, including inter-country (taking into account the EU context), intra-country, intra-cooperative sector and longitudinal comparisons of the evolution of the ACs chosen as case studies, presented a unique, richer approach, going beyond conventional cross-country comparisons.

This book has also made a twofold empirical contribution (Rowe, 2011): First, it has offered a new account of an empirical phenomenon – different expressions of agricultural cooperation – that has challenged existing assumptions about cooperatives. Second, it has revealed something previously undocumented in the literature: the rise of multi-stakeholder cooperatives in European food and farming.

Finally, the evidence provided of different types of strategies aimed at reducing the risk of co-optation currently being implemented by cooperative members in contemporary innovative models has significant societal value. Evidence on these strategies has been captured and theorised, and it is hoped that this research will be useful to the future study and development of fairer and more sustainable food systems. Furthermore, the multivocality approach of this project has also allowed smaller cooperatives and less powerful actors in the food system to voice their opinions and raise issues about trends in the AC sector in their own words.

Implications for food policy and agricultural cooperatives' policy

This research has put forward the concept of cooperative sustainability to suggest that, unless ACs succeed in integrating the four elements of the cooperative triangle, their model will gradually become diluted and no longer transformative. Mechanisms to protect and foster CS include redirecting pressure outwards towards large players in the food system, instead of inwards towards members, and overcoming Western epistemological biases and quantophrenic definitions of success. For this reason, when promoting cooperatives as a way to achieve more sustainable, fairer food systems, we need to be specific about the types of cooperative and cooperation we want to see, and this comes with associated policy implications.

The new analytical framework put forward in Chapter 10 can potentially be used by policymakers as a tool to formulate and assess the impact of policy proposals by analysing how a particular intervention could impact CS, based on the effects/implications for each of the cooperative dimensions proposed. The framework can also be applied to specific cooperatives to study how they have evolved over time and map their relationships with other actors in the food system.

In terms of policy lessons, this research has revealed that an adequate legal framework is not sufficient to encourage MS initiatives, owing to the bottom–up nature of these associations; however, a supportive legal framework can create the right policy context and minimise the diversion of much needed cooperative resources that often go wasted in the process of legally registering an

MSC. The case studies raise questions of whether current CAP subsidies for rural collaboration might be better invested in supporting new models of MS cooperation that are able to foster links between rural and urban development. EU policymakers should consider what kind of cooperation they want to reproduce through subsidies. However, the challenges facing MSCs are not only related to multilevel governance, but also to long-established cultural and social norms and expectations around food, price and consumerist conceptions of progress.

Food democracy and social justice are the underlying issues highlighted by this book (Lang, 2005; Caraher and Dowler, 2014): food democracy not just understood as striving to achieve safe, justly produced and sustainable food for all (Lang, 2005), but democracy in the food policy-making process itself and in the power relations in the food chain. ACs, and especially MSCs, can bring democracy into the equation. Some authors have proposed ACs as a way to counterbalance and reverse the supermarketisation of society (Lawrence and Dixon, 2015), but this is only possible if their vision focuses on relations, rather than on labels, certifications or legal forms that can be easily co-opted and appropriated by the dominant food regime.

At the wider level of national food policy, a bold suggestion emerges from this book. Governments should consider the creation of two parallel food systems in each country: one not-for-profit system for domestic consumption, in an attempt to increase self-reliance, and a second one for export. Currently, there is a mismatch between local consumption and local needs unmet by a global and international food system. Additionally, evidence presented by this research has highlighted a second mismatch, as many ACs have become too large to be able to meet the demand for local food from their communities. Owing to the scale of the operations, they are forced to find larger buyers, decreasing their embeddedness in the rural areas where they operate and their social significance. Bello Horizonte in Brazil is proposed as a well-known case study of a successful food policy programme that is highly reliant on the cooperative model (Sonnino, 2009). Based on the principle of social justice, Bello Horizonte has achieved a better life for its citizens and small-scale farmers through valuing and engaging small and medium ACs that are able to cover local food demand. Successful cooperatives' policies will have to tackle the issue of scale to match local demand. The case studies also showed logistics are a barrier to trans-local or local to local initiatives between like-minded cooperatives based in different countries.

Opportunities for future research

The findings have highlighted areas where further research is required, especially from a sociological perspective, to offer an alternative analysis to the abundant economic literature on ACs. A key suggestion for future research refers to the application of the proposed theoretical framework put forward by this book to individual case studies, including emerging multi-stakeholder models and agricultural workers' cooperatives, to assess differences across the wide cooperative diversity that exists in food and farming throughout different European countries and outside Europe. It would also be of interest to assess potential policy scenarios, exploring the advantages of and barriers to redirecting CAP subsidies from POs to MSCs. An additional under-researched topic involves the evaluation and comparison of farmers' cooperative behaviours in informal versus institutional/legally incorporated cooperative environments.

This book has unravelled the challenge of defining and the risk of quantifying cooperative practices and sustainable diets. Cooperatives, especially open multi-stakeholder cooperatives, are well placed to empower producers and consumers to deal with the complexity of profit imbalances and sustainable diets, in a more democratic and equalitarian way. If any, the most pressing issue, which requires further academic and empirical contributions, concerns researching strategies able to protect the alterity of transformative food and farming initiatives that can realise sustainable food systems. A good start has been made in Chapter 9, but more urgent work is required in this area. No cooperative, organic or fair trade movement is going to achieve its objectives if co-optation and appropriation become an unavoidable fate. The ultimate question emerging from the book is: How can the alterity and the transformative potential of these food and farming movements be protected and fostered in order to achieve fairer and more sustainable food systems for all?

Appendix I General data on agricultural cooperatives in the EU

EU member state	Total no. of cooperatives	Total no. of members	Turnover (€m)
Belgium	301	–	3,257
Bulgaria	900	–	–
Czech Republic	548	524	1,327
Denmark	28	45,710	25,009
Germany	2,400	1,440,600	67,502
Estonia	21	2,036	512
Ireland	75	201,684	14,149
Greece	550	–	711
Spain	3,844	1,179,323	25,696
France	2,400	858,000	84,350
Croatia	613	10,734	167
Italy	5,834	863,323	34,362
Cyprus	14	24,917	62
Latvia	49	–	1,111
Lithuania	402	12,900	714
Luxembourg	55	–	–
Hungary	1,116	31,544	1,058
Malta	18	1,815	204
Netherlands*	215	140,000	32,000
Austria	217	306,300	8,475
Poland	136	–	15,311
Portugal	735	–	2,437
Romania	68	–	204
Slovenia	368	16,539	705
Slovakia	597	–	1,151
Finland	35	170,776	13,225
Sweden	30	160,350	7,438
United Kingdom	200	138,021	6,207
Total	**21,769**	**6,172,746**	**347,342**

Note: * Multiple membership

Source: Copa-Cogeca, 2015.

Appendix II   Why are Spain and the UK two interesting case studies to compare?

Country case	UK	Spain
Differences		
Historical	First country to have a formalised cooperative (Rochdale legacy).	Early agricultural cooperativism (beg. 20th century) became a control tool used by the Church and Fascist regime to counterbalance communist and anarchist ideologies.
	Historical roots of cooperative ideology: – Owenites – Chartism	Historical roots of cooperative ideology: Bakunin and French revolutionary ideas, but high levels of illiteracy and poverty hindered the development of cooperatives.
	First country to undergo industrial revolution and first country to transform its agriculture into a highly industrialised economic sector.	One of the last 17 Eurozone members to industrialise agriculture.
	Second country to consolidate farming (after Denmark; by 1980s UK had the largest farms in Europe).	Concentrated land ownership but still very atomised farming sector.

Sectorial (AC):	Agricultural cooperation developed less and slower than consumer and industrial cooperation.	Agricultural cooperation has a longer tradition than consumer and industrial cooperation.
	Agricultural cooperation became a tool for economic protection and of top–down measures.	Combination of both a tool for economic protection with top–down measures and cooperation as resistance and struggle.
Retail sector	One of the earliest to develop. Highly integrated, reducing farmers' need to find markets for their produce: 367 supermarkets were operating by 1960.	Late development, only 44 stores in 1962.
Coop. policy	Since the Curry report in 2002, government's focus on "collaboration" rather than formal cooperation.	Currently, second strongest co-operative economy in EU.
Farming exports/ imports	UK is a net importer of food, with exports exceeding US$23 billion, compared with imports of around US$54 billion.	Spain is a marginal net exporter: exports of US$33.5 billion vs. US$33 billion imports.

Similarities

Political	Both countries currently have Conservative governments in power promoting cooperatives as part of their economic discourse.	
Legislative	Consolidation of cooperative legislation: In 2014, the Conservative and Lib Dem coalition approved a consolidated Act that brought together 17 different pieces of legislation that had not been updated since 1965.	The Popular Party (Catholic Conservative party) approved a new law in 2013 to foster the integration of agricultural cooperatives and other types of associative agrarian entity.
Economic	Power imbalances and power concentration at producer, processor and retail level created by increasing market concentration and asymmetrical bargaining powers.	

(continued)

(continued)

Country case	UK	Spain
Importance of farming and cooperatives in each country		
Total area (km²)	244,101	505,365
Total population (1,000 inhabitants)	61,176	45,283
Number of holdings (1,000 holdings), 2007	300	1,044
UAA per holding (ha), 2007	53.8	23.8
Employment in agric. (%)	1.4	4.3
GVA/GDP (%)	0.5	2.2
Number of farmers (1,000), 2008 (UK has same numbers in 2008 and 2012)	300	1,160.3
Number of farmers that are members of ACs (including membership in several cooperatives), 2008	150,000	972,380
Number of cooperatives (all sectors), 2008 for Spain and specified for UK	4,820 (2008) 5,933 (2011)	24,738
Number of jobs in cooperatives, 2008	236,000	456,870
Number of jobs in ACs, 2008 for Spain and 2010 for UK	7,950	90,308
Number of ACs	450 (2012 figures) 621 (2016 figures)	3,861
Social economy sector		
Paid employment in cooperatives (number of jobs in 2009–10)	236,000	646,397
Paid employment in the social economy compared with total paid employment in 2009–10 (%)	5.64	6.74

Evolution of paid employment in the social economy from 2002–3 to 2009–10 (%)	−4.57	42.53
Different tax treatment for cooperatives	No	Yes
CIRIEC (International Centre of Research and Information on the Collective Economy) presence?	No	Yes

Source: Compiled by author from several sources. See Chapters 1, 2 and 4 for references.

Appendix III   Summary of interviews by country, type of actor and organisation[1]

Policy triangle	UK (18)	Spain (21.5)	EU/Intl (1.5)
Governance/policymakers			
	• UK policymaker (1)	• Spanish policymaker (1)	
Industry			
Representative bodies	• NFU (1) • The Land Workers' Alliance (1) • Co-operatives UK (1) • Scottish Agricultural Organisation Society (SAOS) (1)	• Cooperativas Agro-alimentarias (1)	• Copa-Cogeca (1)
Agricultural cooperatives	**ACHEIFS**:[2] • Farmway/Mole Valley (1) **New ACs**: • Manchester Veg People (2 + 0.5) • Moss Brook Growers (1) • OrganicLea (1) • Peasant Evolution Producer • Cooperative (1)	**ACHEIFS**: • Anecoop (2) **Traditional cooperatives**: • Valle del Jerte AC (2) • Valle del Tietar AC (3) **New ACs**: • Actyva (2) • Catasol (2) • Cooperative Integral Catalana (1) • Esnetik (3)	

Civil society organisations	• Plunkett Foundation (1) • Somerset Cooperative Services (0.5) • Ecological Land Cooperative (0.5) • Kindling Trust (0.5) • Campaign for Real Farming (2)	• *Soberanía alimentaria, biodiversidad y culturas* magazine (1) • Via Campesina Spanish rep. (also member of an AC) (0.5)	• Via Campesina (Spanish rep.) (0.5)
Academia	• UK academic 1 (1) • UK academic 2 (1)	• Spanish academic (1) • Academic from Finca Experimental University of Almeria–Anecoop (1)	

Notes

1 The numbers in brackets indicate the number of participants interviewed in each category. The number 0.5 is used to indicate cases in which the interviewee was representing two organisations.
2 ACHEIFS refers to ACs highly embedded in industrial food systems.

References

Actyva, 2015. *Cáceres para comérselo: La soberanía alimentaria llevada a la práctica*. Available: www.delaciudadalcampo.net/2015/12/caceres-para-comerselo-la-soberania.html

Ajates Gonzalez, R. 2017a. Going back to go forwards? From multi-stakeholder cooperatives to Open Cooperatives in food and farming. *Journal of Rural Studies*, 53:278–90.

Ajates Gonzalez, R., 2017b. The solution cannot be conventionalised: Protecting the alterity of fairer and more sustainable food networks. In Duncan, J. and Bailey, M. *Sustainable Food Futures: Multidisciplinary solutions*. London: Routledge.

Ajates Gonzalez, R., 2017c. Thank you for the cured meat, but is it grass-fed? In Mata-Codesal, D. and Abranches, M. (eds) *Food Parcels in International Migration. Intimate connections*. Palgrave Macmillan.

Ajates Gonzalez, R., 2017d. Esnetik: Ethics, trust, transparency and the challenges of negotiating meaningful sustainability. In Wakeford, T., Chang, M. and Anderson, C. *Action Research for Food Systems Transformation*. Centre for Agroecology, Water and Resilience, Coventry University.

Allen, P., 2004. *Together at the Table: Sustainability and sustenance in the American agrifood system*. University Park, PA: Penn State Press.

Alonso Mielgo, A. M. and Casado, G., 2000. Asociaciones de productores y consumidores de productos ecologicos en Andalucia: una experiencia de canales cortos de distribucion de productos de calidad. *Phytoma España*, 124:30–8.

Alter-EU, 2012. Who's driving the agenda at DG Enterprise and Industry? The dominance of corporate lobbyists in DG Enterprise's expert groups. Available: corporateeurope.org/sites/default/files/dgentr-driving.pdf

Altieri, M., 2007. Fatal Harvest: Old and new dimensions of the ecological tragedy of modern agriculture. In: Nemetz, P. (ed.) *Sustainable Resource Management: Reality or illusion*, pp. 189–213. Cheltenham: Edward Elgar.

Anderson, C. R., Brushett, L., Gray, T. W. and Renting, H., 2014. Working together to build cooperative food systems. *Journal of Agriculture, Food Systems & Community Development*, 4(3):3–9.

Anecoop, 2013. *Annual report 2012/2013*. Available: http://static-01.simauria.com/store/anecoop/memoria_anecoop_2013.pdf

Anecoop, 2015. *Memoria de Responsabilidad Social Corporativa 2014/2015*. Available: http://anecoop.com/wp-content/uploads/2016/03/memoriaANECOOPok2016.pdf

Anecoop, 2016. *Vision Anecoop*. Available: http://anecoop.com/sobre-nosotros/filosofia/

Anglia Farmers, 2016. *Driving the future of farming through traditional values*. Available: www.angliafarmers.co.uk/about/

Arthur, L., Keenoy, T., Scott Cato, M. and Smith, R., 2008. *Social movements and contention: An exploration of the implications of diachronic and synchronic change.* Transborder Laboratories from Below Seminar Proceedings. Brno: Economy and Society Trust and Institute for Studies in Political Economy.

Axelrod, R. and Hamilton, W.D., 1981. The evolution of cooperation. *Science*, 211(4489):1390–6.

Bachmann-Medick, D., 1996. Cultural misunderstanding in translation: Multicultural coexistence and multicultural conceptions of world literature. *Erfurt Electronic Studies in English*, 7:7–96.

Baggini, J., 2014. *The Virtues of the Table: How to eat and think.* London: Granta Books.

Baranchenko, Y. and Oglethorpe, D., 2012. The potential environmental benefits of co-operative businesses within the climate change agenda. *Business Strategy & the Environment*, 21(3):197–210.

Barke, M. and Eden, J., 2001. Co-operatives in southern Spain: Their development in the rural tourism sector in Andalucía. *International Journal of Tourism Research*, 3(3):199–210.

Bartlett, W. and Pridham, G., 1991. Co-operative enterprises in Italy, Portugal and Spain: History, development and prospects. *Journal of Interdisciplinary Economics*, 4(1):33–59.

Barton, D., 1989. What is a cooperative? In Cobia, D. (ed.) *Cooperatives in Agriculture*, pp. 1–19. Englewood Cliffs, NJ: Prentice-Hall.

Bauwens, M., 2014. *Michel Bauwens on the rise of multi-stakeholder cooperatives.* Peer to Peer blog. Available: http://blog.p2pfoundation.net/michel-bauwens-onthe-rise-of-multi-stakeholder-cooperatives/2014/11/13

Bauwens, M. and Kostakis, V., 2014. From the communism of capital to capital for the commons: Towards an open co-operativism. *TripleC, Communication, Capitalism & Critique, Open Access Journal for a Global Sustainable Information Society*, 12(1):356–61.

Baxter, P. and Jack, S., 2008. Qualitative case study methodology: Study design and implementation for novice researchers. *The Qualitative Report*, 13(4):544–59.

BBC, 2013. *Q&A: Reform of EU farm policy.* Available: www.bbc.co.uk/news/world-europe-11216061

Beck, U., 1992. *Risk Society: Towards a new modernity.* London: Sage.

Becker, H. S., 1998. *Tricks of the Trade.* Chicago: University of Chicago Press.

Berger, P. L., and Luckerman, T. 1966. *The Social Construction of Reality: A treatise in the sociology of knowledge.* London: Penguin.

Bernal, A., Heras, M. O. and Villaverde, A. L. L., 2001. *Entre surcos y arados: el asociacionismo agrario en la España del siglo 20th.* Cuenca, Spain: Universidad de Castilla La Mancha.

Berthelot, J., 2012. The European agricultural cooperatives, promoters of the unequal globalization. Solidarité. Available: http://solidarite.asso.fr/Papers-2012

Bhabha, H. K., 1994. *The Location of Culture.* Abingdon, UK: Routledge.

Bijman, J., Lindgreen, A., Hingley, M. K., Harness, D., and Custance, P., 2010. Agricultural cooperatives and market orientation: A challenging combination? In Lindgreen, A., Hingley, M. K., Harness, D. and Custance, P. (eds) *Market Orientation: Transforming food and agribusiness around the customer*, pp. 119–36. Surrey: Gower.

Bijman, J., Iliopoulos, C.. Poppe, K. et al., 2012. *Support for farmers' cooperatives, Case study report.* Wageningen: Wageningen UR. Available: www.wageningenur.nl/en/show/Support-for-Farmers-Cooperatives.htm

Birchall, J., 1994. *Co-op: The people's business.* Manchester: Manchester University Press.

Birchall, J., 2003. *Rediscovering the Cooperative Advantage: Poverty reduction through self-help.* Geneva: International Labour Organization.

Birchall, J., 2005. *Co-operative Principles Ten Years On*. International Co-operative Alliance, Issue 2, 98(2):45–63.

Birchall, J., 2010. *People-Centred Businesses: Co-operatives, mutuals and the idea of membership*. London: Palgrave Macmillan.

Birchall, J., 2012. *Sustainable Development of Agricultural Cooperatives*. Presentation at ICAO General Meeting, 26 November 2012, Kobe, Japan.

Blatchford, R., 1902. *Britain for the British*. London: Clarion Press.

Bobel, C., 2007. "I'm not an activist, though I've done a lot of it": Doing activism, being activist and the "perfect standard" in a contemporary movement. *Social Movement Studies*, 6(2):147–59.

BOE, 2013. Ley 13/2013 de Fomento de la integración de cooperativas y de otras entidades asociativas de carácter agroalimentario. Madrid: Boletín Oficial del Estado.

Böhm, S., 2006. *Repositioning Organization Theory: Impossibilities and strategies*. London: Palgrave.

Böhm, S. and Land, C., 2009. No measure for culture? Value in the new economy. *Capital & Class*, 97:75–98.

Böhm, S., Dinerstein, A. C. and Spicer, A., 2010. (Im)possibilities of autonomy: Social movements in and beyond capital, the state and development. *Social Movement Studies*, 9(1):17–32.

Böhm, S., Pervez Bharucha, Z. and Pretty, J. 2014. *Ecocultures: Blueprints for sustainable communities*. Abingdon, UK: Routledge.

Borzaga, C., Depedri, S. and Tortia, E. C., 2010. The growth of organizational variety in market economies: The case of social enterprises. Euricse Working Papers, n.003/10.

Brazda, J. and Schediwy, R., 1989. *Consumer Co-operatives in a Changing World*. Geneva: ICA.

Brunori, G., Rossi, A. and Guidi, F., 2012. On the new social relations around and beyond food. Analysing consumers' role and action in Gruppi di Acquisto Solidale (Solidarity Purchasing Groups). *Sociologia Ruralis*, 52(1):1–30.

Bryman, A., 2003. *Quantity and Quality in Social Research*. London: Routledge.

BTC, 2014. *New fair trade labels attract criticisms*. Available: www.befair.be/en/content/new-fairtrade-labels-attract-criticism

Buck, D. et al., 1997. From farm to table: The organic vegetable commodity chain of northern California. *Sociologia Ruralis*, 37(1):3–20.

Burge, A., 2012. Looking back towards a Welsh co-operative future? (Background paper on the Welsh Government Commission on Co-operatives and Mutuals). Available: http://alunburgeassociates.coop/wp-content/uploads/2012/12/Look-Back-Towards-a-Welsh-Co-operative-Future-Background-Paper-for-Welsh-Government-Commission.pdf

Burnett, J., 1985. *Plenty and Want: A social history of food in England from 1815 to the present day*, 4th edn. New York: Methuen.

Cairney, P., 2012. *Understanding Public Policy, Theories and Issues*. Basingstoke: Palgrave Macmillan.

Calle Collado, A. and Gallar, D., 2010. Agroecología Política: Transición Social y Campesinado. VIII Rural Sociology Latin American Congress, 15–19 November 2010, Group 2. Porto de Galinhas, Pernambuco, Brazil.

Calle Collado, A., Soler Montiel, M. et al., 2012. La desafección al sistema agroalimentario: ciudadanía y redes sociales. *Interface: A Journal for and about Social Movements*, 4(2):459–89.

Campos, V. and Carreras, L., 2012. Situación económica y financiera de las cooperativas hortofrutícolas catalanas. Estudio empírico aplicado a la provincia de Tarragona. *CIRIEC-España, Revista de Economía Pública, Social y Cooperativa*, 74:149–76.

Caraher, M. and Dowler, E., 2007. Food projects in London: Lessons for policy and practice – A hidden sector and the need for more unhealthy puddings . . . sometimes. *Health Education Journal*, 66(2):188–205.

Caraher, M. and Dowler, E., 2014. Food for poorer people: Conventional and "alternative" transgressions. In Goodman, M. and Sage, C. (eds) *Food Transgressions: Making sense of contemporary food politics*, pp. 227–46. Farnham, UK: Ashgate.

Caraher, M., Smith, J. and Machell, G., 2014. To co-operate or not co-operate: A case study of food co-ops in England. *Journal of Co-operative Studies*, 47(2): 6–19.

Cartwright, A. and Swain, N., 2003. Dividing the rural sector: Finding farmers in Central and Eastern Europe. *Rural Transition Series*, Working Paper no. 53. Liverpool: University of Liverpool, Centre for Central and Eastern European Studies.

CASI, 2015. *Actyva*. Available: www.casi2020.eu/casipedia/cases/actyva/

CECOP, 2013. The European Parliament pinpoints the contribution of cooperatives to overcoming the crisis. Available: www.cecop.coop/The-European-Parliament-approves-a

Cervantes, C. R. and Fernandes, M., 2008. Cooperativism in Portugal and Spain from the dictatorships to European integration, 1940–2006. IX Spanish Association of Economic History Congress, University of Murcia, Spain.

Chaddad, F., 2009. Both markets and hierarchy: Understanding the hybrid nature of cooperatives. *International Workshop on Rural Cooperation in the 21st Century: Lessons from the Past, Pathways to the Future*. Rehovot, Israel, 15–17 June 2009.

Chapagain, A. K. and Hoekstra, A. Y., 2006. Water footprints of nations, vols 1 & 2. *UNESCO-IHE Value of Water Research Report Series*, no. 16. Paris: UNESCO.

Chatterton, P. and Pickerill, J., 2010. Everyday activism and transitions towards post-capitalist worlds. *Transactions of the Institute of British Geographers*, 35(4): 475–90.

Chile Alimentos, 2015. *Tomates, España: Estudio indica que El agua desalada produce 4 kilos más de tomate por metro que la de pozo*. Available: www.chilealimentos.com/2013/index.php/es/noticias/alimentos-procesados/conservas/22657-tomates-españa-estudio-indica-que-el-agua-desalada-produce-4-kilos-más-de-tomate-por-metro-que-la-de-pozo.html

Chloupkova, J., Svendsen, G. L. H. and Svendsen, G. T., 2003. Building and destroying social capital: The case of cooperative movements in Denmark and Poland. *Agriculture & Human values*, 20(3):241–52.

Clancy, K. and Lockeretz, W., 1997. Reconnecting farmers and citizens in the food system. In Lockeretz, W. (ed.) *Visions of American Agriculture*, pp. 47–57. Ames, IA: Iowa State University.

Clark, J., 2006. The institutional limits to multifunctional agriculture: Subnational governance and regional systems of innovation. *Environment & Planning C: Government & Policy*, 24(3):331–49.

Clark, J., and Jones, A., 2011. Telling stories about politics: Europeanization and the EU's Council Working Groups. *JCMS: Journal of Common Market Studies*, 49(2):341–66.

CNC, 2012. IPN 82/12. *Anteproyecto de Ley de Fomento de la Integración Cooperativa y Asociativa*. Comisión Nacional de los Mercados y la Competencia. Available: www.cnmc.es/es-es/promoción/informessobrenormativa.aspx?num=IPN+082%2F13&ambito=Informes+de+Propuestas+Normativas&b=&p=86&ambitos=Informes+de+Propuestas+Normativas&estado=0&sector=0&av=0

Coates, K. and Benn, T., 1976. *The New Worker Co-operatives*. Nottingham: Spokesman Books for the Institute for Workers' Control.

COCERAL, 2015a. Review of the EU's decision-making process to authorise GMOs for food and feed uses. Position paper, 30.3.15. Brussels: COCERAL.

COCERAL, 2015b. *EU food and feed chain partners reject EU Commission proposal which threatens internal market for agri-food products.* Position paper, 22.4.15. Available: www.europabio.org/sites/default/files/press/pr_review_gm_authorisation_food_feed_uses_22_april_2015.pdf

COCETA, 2012. El Congreso de los Diputados conmemora el Año Internacional de las Cooperativas de las Naciones Unidas. Available: www.coceta.coop/noticias-coceta.asp?idnew=322

Cogeca, 2010. *Agricultural cooperatives in Europe, main issues and trends.* Available: www.agro-alimentarias.coop/ficheros/doc/03020.pdf

Cogeca, 2015. *Development of agricultural cooperatives in the EU 2014.* Brussels: Cogeca.

Conaty, P., 2015. *A cooperative economy for the common good.* Wales Co-operative Centre. Available:http://wales.coop/file/A-Collaborative-Economy-for-the-Common-Good.pdf

Cook, M. L., 1995. The future of US agricultural cooperatives: A neo-institutional approach. *American Journal of Agricultural Economics*, 77(5):1153–9.

Cook, M. L. and Iliopoulos, C., 1999. Beginning to inform the theory of the cooperative firm: Emergence of the new generation cooperative. *The Finnish Journal of Business Economics Publications*, 4:525–35.

Cook, M. L., Iliopoulos, C. and Chaddad, F. R., 2004. Advances in cooperative theory since 1990: A review of agricultural economics literature. In Hendrikse, G. W. J. (ed.) *Restructuring Agricultural Cooperatives*, pp. 65–90. Rotterdam: Haveka.

Cook, M. L. and Plunkett, B., 2006. Collective entrepreneurship: An emerging phenomenon in producer-owned organizations. *Journal of Agricultural & Applied Economics*, 38(2):421–8.

Cook, T. D., 1985. Post-positivist critical multiplism. In *Evaluation Studies Review Annual*. Oxford: Sage.

Cooperativa Integral Catalana, 2016. *General principles.* Available: http://cooperativa.cat/en/whats-cic/general-principles/

Cooperativas Agro-alimentarias, 2013. Estudio diagnóstico sobre la situación del Cooperativismo Agroalimentario en España, Análisis DAFO. Madrid: Cooperativas Agro-alimentarias.

Cooperativas Agro-alimentarias, 2016. *Qué es Cooperativas Agro-alimentarias.* Available: www.agro-alimentarias.coop/informacion_corporativa

Co-operative Group, 2014. *The Co-operative Group agrees to sale of farms business for £249 million to the Wellcome Trust.* Available: www.co-operative.coop/corporate/press/press-releases/headline-news/farms-sale/

Cooperative Heritage Trust, 2012. *Our story – Rochdale Pioneers Museum.* Available: www.rochdalepioneersmuseum.coop/wp-content/uploads/2013/02/Our-Story.pdf

Co-operative Party, 2014. *About. History.* Available: https://party.coop/about/history

Cooperatives Europe, 2016. *The power of cooperation. Cooperatives Europe key figures 2015.* Available: https://coopseurope.coop/sites/default/files/The%20power%20of%20Cooperation%20-%20Cooperatives%20Europe%20key%20statistics%202015.pdf

Co-operatives UK, 2011. *Facts and figures.* Available: www.uk.co-op/economy/figures

Co-operatives UK, 2012. *The co-operative economy in Wales 2012.* Available: www.cooperatives-wales.coop/wp-content/uploads/2009/11/Brochure-2012.pdf

Co-operatives UK, 2013a. *Corporate governance code for agricultural co-operatives*. Available: www.uk.coop/sites/default/files/uploads/attachments/agricultural_govern ance_code_edit_0.pdf

Co-operatives UK, 2013b. *Homegrown. The UK co-operative economy 2013*. Available: http://ica.coop/en/media/library/member-publication/homegrown-uk-co-operative-economy-2013

Co-operatives UK, 2014a. *Agricultural co-operatives*, Policy Position Briefing. Manchester: Co-operatives UK.

Co-operatives UK, 2014b. *Withdrawable share capital increase: What does this mean for co-operatives?* Available: www.thenews.coop/46776/news/general/withdrawable-share-capital-increase-mean-co-operatives/

Co-operatives UK, 2016. Agricultural co-operatives Report on the co-operative farming sector. Available: www.uk.coop/sites/default/files/uploads/attachments/report_on_the_co-operative_agriculture_sector_-_2016.pdf

Co-operatives UK, 2017. *Reimagine the economy: The UK co-operative economy 2017*. Available: http://reports.uk.coop/economy2017/

Copa, 2016. Copa: European Agricultural Union. Available: www.copa-cogeca.be/CopaHistory.aspx

Copa-Cogeca, 2009. Fruit and vegetable producer organisations in the EU: Overview and prospects. Available: www.Copa-Cogeca.be/img/user/7493_E.pdf

Copa-Cogeca, 2011. Future challenges for sustainable production from a farmers' point of view. *Transition Towards Sustainable Food Consumption and Production in a Resource Constrained World Conference*, 5 May, Budapest.

Copa-Cogeca, 2012. *The Common Agricultural Policy after 2013*. Available: www.Copa-Cogeca.be/img/user/file/PAC2013/2012futureCAP_EN.pdf

Copa-Cogeca, 2015. *Development of agricultural cooperatives in the EU 2014*. Available: http://zadruge.coop/upload_data/site_files/development-of-agricultural-cooperatives-in-the-eu_2014.pdf

Corporate Europe, 2014. *Will public trust in the EU be sacrificed to keep agribusiness happy?* Available: http://corporateeurope.org/expert-groups/2014/01/will-public-trust-eu-be-sacrificed-keep-agribusiness-happy

Cordell, D., Drangert, J. O. and White, S., 2009. The story of phosphorus: Global food security and food for thought. *Global Environmental Change*, 19(2):292–305.

Cox, R., Kneafsey, M., Holloway, L. et al., 2013. Greater than the sum of the parts? Unpacking ethics of care within a community supported agriculture scheme. In Goodman, M. K. (ed.) *Food Transgressions: Making sense of contemporary food politics*. Critical Food Studies. Aldershot, UK: Ashgate Press.

Cultivate Oxford, 2016. *Cultivate is community owned*. Available: http://cultivateoxford.org

Curry, D., 2002. *Farming and food: A sustainable future*. Report of the Policy Commission on the Future of Farming and Food. London: Cabinet Office.

Darwin, C., 1971. *The Descent of Man, and Selection in Relation to Sex*. New York: D. Appleton.

Davey, B., Josling, T. E. and McFarquhar, A., 1976. *Agriculture and the State. British policy in a world context*. London: Macmillan.

DEFRA, 2012. *Food Statistics Pocket Book 2012*. London: Department for Environment, Food and Rural Affairs. Available: http://webarchive.nationalarchives.gov.uk/20130123162956/http://www.defra.gov.uk/statistics/files/defra-stats-food farm-food-pocketbook-2012-130104.pdf

DEFRA, 2015. *Agriculture in the United Kingdom 2015.* Available: www.gov.uk/government/uploads/system/uploads/attachment_data/file/535996/AUK-2015-07jul16.pdf

DEFRA, 2016. *Corporate report – Single departmental plan: 2015 to 2020.* Available: www.gov.uk/government/publications/defra-single-departmental-plan-2015-to-2020/single-departmental-plan-2015-to-2020

De Schutter, O., 2014. *The power of procurement.* Briefing note. Available: www.srfood.org/images/stories/pdf/otherdocuments/20140514_procurement_en.pdf

Denzin, N. K. and Giardina, M. D., 2008. *Qualitative Inquiry and the Politics of Evidence.* Walnut Creek, CA: Left Coast Press.

Denzin, N. K. and Lincoln, Y. S. (eds), 1998. *Strategies of Qualitative Inquiry.* London: Sage.

Drewnowski, A., 2017. Sustainable, healthy diets: Models and measures. In *Sustainable Nutrition in a Changing World*, pp. 25–34. Cham: Springer.

Dunn, J. R., 1988. Basic cooperative principles and their relationship to selected practices. *Journal of Agricultural Cooperation*, 3:83–93.

Duran, E., 2016. The activist who stole over half a million dollars to fund a new anti-capitalist movement – Interview with Enric Duran. Available: www.vice.com/video/daily-vice-robin-bank-enric-duran

Ecological Land Cooperative, 2015. *Our 2015–2020 business plan.* Available: http://ecologicalland.coop/business-plan

EFFP, 2004. *Farming and Food: Collaborating for profit.* London: English Food and Farming Partnership.

EFFP, 2013. *Cooperatives and Producer Organisations Survey.* Available: www.surveymonkey.com/r/cooperatives_and_producer_organisations?sm=Kl7Bf4nDX8U1E ZWTxiFfiIY1TkfvJffO3feof%2bO460Y%3d

EFFP, 2014. Conditions, Attitudes and Structures of Successful POs and Cooperatives. London: English Food and Farming Partnership

Egdell, J. M. and Thomson, K. J., 1999. The influence of UK NGOs on the common agricultural policy. *JCMS: Journal of Common Market Studies*, 37(1):121–31.

Ellingson, L. L., 2009. *Engaging Crystallization in Qualitative Research: An introduction.* London: Sage.

El Pais, 1986, Creada la Asociación Española de Cooperativas Agrarias. *El Pais*, 14 May 1986. Available: http://elpais.com/diario/1986/05/14/economia/516405615_850215.html

Emery, S. B., 2015. Independence and individualism: Conflated values in farmer cooperation? *Agriculture & Human Values*, 32(1):47–61.

Emery, S. B., Forney, J. and Wynne-Jones, S., 2017. The more-than-economic dimensions of cooperation in food production. *Journal of Rural Studies*, 53:229–35.

Encina Duval, B. et al., 2011. Las cooperativas hortofrutícolas frente a la crisis. La necesaria apuesta por la competitividad. Aspectos económico-financieros. *CIRIEC-España, Revista de economía pública, social y cooperativa*, 72:125–56.

Environment, Food and Rural Affairs Committee, 2010. *Dairy farmers of Britain*, Fifth Report of Session 2009–10, vol. 2. London: House of Commons.

ESCA, 1986. *The Cooperative, Mutual and Non-Profit Sector and its Organizations in the European Community.* Luxembourg: Economic and Social Consultative Assembly, Office for Official Publications of the European Communities.

Esnetik, 2015. *¿Qué es Esnetik?* Available: www.esnetik.com/?page_id=14

EU Joint Research Centre, 2008. *JRC Workshop on Global Commercial Pipeline of New GM Crops.* Seville, 12–13 November. Available: http://igtcglobal.org/filead min/user_upload/JRC_GM_pipeline_workshop_in_Seville__Minutes.pdf

Europa Press, 2015. La superficie de invernaderos crece un 10,5% en los últimos cuatro años hasta llegar a las 29.596 hectáreas. *Europa Press.* Available: www.europapress. es/andalucia/almeria-00350/noticia-superficie-invernaderos-crece-105-ul timos-cuatro-anos-llegar-29596-hectareas-20150213102204.html

European Commission, 2012a. *The Common Agricultural Policy: A story to be continued.* Available: http://ec.europa.eu/agriculture/50-years-of-cap/files/history/history_book_ lr_en.pdf

European Commission, 2012b. *National accounts and GDP.* Available: http://epp.euro stat.ec.europa.eu/statistics_explained/index.php/National_accounts_-_GDP

European Commission, 2013a. *How many people work in agriculture in the European Union?* Available: http://ec.europa.eu/agriculture/rural-area-economics/briefs/pdf/ 08_en.pdf

European Commission, 2013b. *The history of the CAP.* Available: http://ec.europa.eu/ agriculture/cap-history/

European Commission, 2013c. *Common Market Organisation.* Available: http://ec. europa.eu/enterprise/glossary/cmo_en.htm

European Commission, 2014a. *The economic impact of modern retail on choice and innova-tion in the EU food sector.* Luxembourg: Publications Office of the European Union.

European Commission, 2014b. *Assessing efficiencies generated by agricultural producer organi-sations.* Luxembourg: Publications Office of the European Union.

European Commission, 2014c. *Report on the fruit and vegetables regime.* Available: www.europarl.europa.eu/meetdocs/2014_2019/documents/com/com_com (2014)0112_/com_com(2014)0112_en.pdf

European Union, 2011. *Structural development in EU agriculture.* Agricultural Economic Briefs, Brief no. 3. Available: http://ec.europa.eu/agriculture/rural-area-economics/ briefs/pdf/03_en.pdf

European Union, 2012. *Eurostat Pocketbooks: Agriculture, fishery and forestry statistics,* 2012 edn. Luxembourg: European Union Publications.

Eurostat, 2012. Agricultural census in Spain. Available: http://ec.europa.eu/eurostat/ statistics-explained/index.php/Agricultural_census_in_Spain

Fairbairn, B., 1994. *The Meaning of Rochdale: The Rochdale pioneers and the co-operative prin-ciples.* Canada: University of Saskatchewan, Centre for the Study of Co-operatives.

Fairbairn, B., 2004. Self-help and philanthropy: The emergence of cooperatives in Britain, Germany, the United States, and Canada from mid-nineteenth to mid-twentieth century. In Adam, T. (ed.) *Philanthropy, Patronage, and Civil Society: Experiences from Germany, Great Britain, and North America.* Bloomington, IN: Indiana University Press.

Fairbairn, B., 2006. *Cohesion, Adhesion, and Identities in Co-operatives.* Canada: University of Saskatchewan, Centre for the Study of Co-operatives.

Fairlie, S. 2007. Can Britain feed itself? *The Land,* 4:2007–8.

Fairtrade Foundation, 2011. *Fair and Local: Farmers of the world unite?* London: Fairtrade Foundation.

FAO, 1995. *Dimensions of Need: An atlas of food and agriculture.* Rome: Food and Agriculture Organisation.

FAO, 1998. *Women: Users, preservers, and managers of agro-biodiversity.* Rome: Food and Agriculture Organisation.

FAO, 2010a. *Sustainable diets and biodiversity*. Rome: Food and Agriculture Organisation Available: www.fao.org/docrep/016/i3004e/i3004e.pdf

FAO, 2010b. Biodiversity International. *Final document: International Scientific Symposium: Biodiversity and Sustainable Diets, United against Hunger*. 3–5 November. Rome: Food and Agriculture Organisation.

FAO, 2012. *Agricultural Cooperatives: Key to feeding the world*. Rome: Food and Agriculture Organization of the United Nations.

Favaro, L., 2017. Mediating intimacy online: Authenticity, magazines and chasing the clicks. *Journal of Gender Studies*, 26(3):321–34.

Feindt, P. H. and Flynn, A., 2009. Policy stretching and institutional layering: British food policy between security, safety, quality, health and climate change. *British Politics*, 4(3):386–414.

Feng, L., Nilsson, J., Ollila, P. and Karantininis, K., 2011. The human values behind farmers' loyalty to their cooperatives. *5th International Conference on Economics and Management of Networks*. 1–3 December, Limassol, Cyprus.

Ferris. J. L., 2017. Data privacy and protection in the agriculture industry: Is federal regulation necessary? *Minnesota Journal of Law, Science & Technology*, 18(1). Available: https://scholarship.law.umn.edu/cgi/viewcontent.cgi?article=1422&context=mjlst

Finlay, L., 2002. Negotiating the swamp: The opportunity and challenge of reflexivity in research practice. *Qualitative Research*, 2(2):209–30.

Foco Sur, 2011. *Centinelas de la biodiversidad*. Available: www.gem.es/areas_trabajo/articulos/r_semillas.pdf

Foley, J. A., Ramankutty, N., Brauman, K. A. et al., 2011. Solutions for a cultivated planet. *Nature*, 478(7369):337–42.

Folke, C., Jansson, Å., Larsson, J., Costanza, R., 1997. Ecosystem appropriation of cities. *Ambio*, 26(3):167–72.

Fonte, M. and Cucco, I., 2017. Cooperatives and alternative food networks in Italy. The long road towards a social economy in agriculture. *Journal of Rural Studies*, 53:291–302.

Forney, J. and Häberli, I., 2017. Co-operative values beyond hybridity: The case of farmers' organisations in the Swiss dairy sector. *Journal of Rural Studies*, 53:236–46.

Food Drink Europe & Copa-Cogeca, 2014. *Annex key non-tariff measures*. Available: www.Copa-Cogeca.be/Download.ashx?ID=1130938

Food from Britain (no date). Food from Britain has closed. Available: www.gov.uk/government/organisations/food-from-britain

Frazier, J., 1997. Sustainable development: Modern elixir or sack dress? *Environmental Conservation*, 24(02):182–93.

Fresh Plaza, 2016. *El "software" que está mejorando resultados en las empresas agrícolas*. Available: www.freshplaza.es/article/96122/El-software-que-está-mejorando-resultados-en-las-empresas-agr%C3%ADcolas

Friedmann, H., 2005. Feeding the Empire: The pathologies of globalized agriculture. *Socialist Register*, 41:125–43.

Fromartz, S., 2007. *Organic, Inc.: Natural foods and how they grew*. Florida: Houghton Mifflin Harcourt.

Fuller, D., Jonas, A. E. and Lee, R. (eds), 2010. *Interrogating Alterity: Alternative economic and political spaces*. Farnham, UK: Ashgate.

Fulton, M., 2001. *Traditional versus New Generation Cooperatives: A cooperative approach to local economic development*. Westport, CT: Quorum Books.

Fundacion Finca Experimental, 2013. *Memoria de Actividades 2012–2013*. Almeria: Fundacion Universidad de Almeria – Anecoop.

Galdeano-Gómez, E., Pérez-Mesa, J. C. and Giagnocavo, C. L., 2015. Food exporters and co-opetition relationships: An analysis on the vegetable supply chain. *British Food Journal*, 117(5):1596–609.

Gall, P. A., 1993. Policy paradigms, social learning and the state. *Comparative Politics*, 25:275–96.

Garnett, T., 2014. Three perspectives on sustainable food security: Efficiency, demand restraint, food system transformation. What role for life cycle assessment? *Journal of Cleaner Production*, 73:10–18.

Garnett, T., 2016. Plating up solutions. *Science*, 353(6305):1202–4.

Garnett, T., Appleby, M. C., Balmford, A. et al., 2013. Sustainable intensification in agriculture: Premises and policies. *Science*, 341(6141):33–4.

Garnett, T. and Godfray, C., 2012. *Sustainable Intensification in Agriculture: Navigating a course through competing food system priorities*. Food Climate Research Network and the Oxford Martin Programme on the Future of Food. Oxford: University of Oxford.

Garrido Herrero, S., 2003. El primer cooperativismo agrario español. *Ciriec-España*, 44:33–56.

Garrido Herrero, S., 2007. Why did most cooperatives fail? Spanish agricultural cooperation in the early twentieth century. *Rural History*, 18:183–200.

Gentilini, U., 2013. Banking on food: The state of food banks in high-income countries. *IDS Working Papers*, 415:1–18.

GeoSAS, 2012. Rwanda consultation report. African Gender, Climate Change and Agriculture Support Program (GCCASP). Addis Ababa: African Union.

Gerber, P. J., Steinfeld, H., Henderson, B. et al., 2013. *Tackling Climate Change through Livestock – A global assessment of emissions and mitigation opportunities*. Rome: Food and Agriculture Organization of the United Nations.

GESTHA, 2014. *La economía sumergida pasa factura*. Available: www.gestha.es/archivos/actualidad/2014/2014-01-29_InformePrensa_EconomiaSumergida.pdf

Giagnocavo, C., 2012. *The Almería Agricultural Cooperative Model: Creating successful economic and social communities*. New York: United Nations, Division for Social Policy and Development, Department of Economic and Social Affairs.

Giagnocavo, C. and Vargas-Vasserot, C., 2012. *Support for farmers' cooperatives; Country report Spain*. Wageningen: Wageningen UR.

Gil, E. I. M., 2004. Falangistas y católicos sociales en liza por el control de las cooperativas. *Historia del presente*, 3:29–43.

Glasman, M., 1996. *Unnecessary Suffering: Managing markets utopia*. London: Verso.

Goldsmith, P. D. and Kane, S., 2002. *The farm business environment and new generation cooperatives as an innovation strategy*. Paper presented at NCR-194 Research on Cooperatives Annual Meeting, 13 November, St Louis, MO.

Gómez López, J. D., 2004. *Las cooperativas agrarias. Instrumento de desarrollo rural*. Alicante: Universidad de Alicante.

Gómez López, J. D., 2009. El movimiento cooperativo agrario en España y la Unión Europea: Evolución y cambios verificados ante el proceso de internacionalización del capital. *Boletín de Geografía*:15–23.

Goodman, D. and DuPuis, E. M., 2002. Knowing food and growing food: Beyond the production consumption debate in the sociology of agriculture. *Sociologia Ruralis*, 42(1):5–22.

Goodman, D., DuPuis, E. M. and Goodman, M. K., 2011. *Alternative Food Networks: Knowledge, practice, and politics*. Abingdon, UK: Routledge.

Gray, T. W., Heffernan, W. D. and Hendrickson, M. K., 2001. Agricultural cooperatives and dilemmas of survival. *Journal of Rural Cooperation*, 29(2):167–92.

Gray, T. W. and Stevenson, G. W., 2008. Cooperative structure for the middle: mobilizing for power and identity. In Lyson, T. A., Stevenson, G. W. and Welsh, R. (eds) *Food and the Mid-level Farm: Renewing an agriculture of the middle*, pp. 37–53. Cambridge: MIT Press.

Gray, T. W., 2014a. Agricultural cooperatives. In Thompson, P. and Kaplan, D. (eds) *Encyclopedia of Food and Agricultural Ethics*, pp. 46–54. Dordrecht, Netherlands: Springer.

Gray, T. W., 2014b. Historical tensions, institutionalization, and the need for multi-stakeholder cooperatives. *Journal of Agriculture, Food Systems, & Community Development*, 4(3):23–8.

Grenfell, M., 1998. Border-crossing: Cultural hybridity and the rural and small schools practicum. Annual Conference of the Australian Association for Research in Education. Nov–Dec 1998, Adelaide, Australia.

Griffiths, P., 2012. Ethical objections to fairtrade. *Journal of Business Ethics*, 105(3):357–73.

Guerin, J. R., 1964. Limitations of supermarkets in Spain. *The Journal of Marketing*, 28:22–6.

Guinnane, T. W. and Martínez-Rodríguez, S., 2010. Did the cooperative start life as joint-stock company? Business law and cooperatives in Spain, 1869–1931. Yale University, Center discussion paper no. 987.

Gussow, J. and Clancy, K. L., 1986. Dietary guidelines for sustainability. *Journal of Nutrition Education*, 18(1):1–5.

Gutierrez, K. D., Baquedano-López, P. and Tejeda, C., 1999. Rethinking diversity: Hybridity and hybrid language practices in the third space. *Mind, Culture, & Activity*, 6(4):286–303.

Guzmán, G. G. and Mielgo, A. 2007. La investigación participativa en agroecología: Una herramienta para el desarrollo sustentable. *Revista Ecosistemas*, 16(1).

Hall, P. A., 1993. Policy paradigms, social learning, and the state: The case of economic policymaking in Britain. *Comparative Politics*, 25(3):275–96.

Hanisch, M. and Rommel, J., 2012. Support for farmers' cooperatives; Case study report. Producer organizations in European dairy farming. Wageningen: Wageningen UR.

Hantrais, L., 1999. Contextualization in cross-national comparative research. *International Journal of Social Research Methodology*, 2(2):93–108.

Harris, A., Stefanson, B. and Fulton, M., 1996. New generation cooperatives and cooperative theory, *Journal of Cooperatives*, 11:15–28.

Helmberger, P. and Hoos, S., 1962. Cooperative enterprise and organization theory. *Journal of Farm Economics*, 44(2):275–90.

Henrich, J. and Henrich, N., 2007. *Why Humans Cooperate: A cultural and evolutionary explanation*. Oxford: Oxford University Press.

Henrÿ, H., 2012. *ILO guidelines on co-operative legislation*, 3rd revised edn. International Labour Organization. Available: http://ica.coop/en/media/news/ilo-updates-guidelines-co-operative-legislation

Hermi Zaar, M., 2010. El movimiento cooperativo agrario en España durante la segunda mitad del siglo 19th y primer tercio del siglo 20th. *Biblio 3w: Revista Bibliográfica de Geografía y Ciencias Sociales*, 15(868).

Hertzler, J. O., 1931. Communist and Co-operative Colonies by Charles Gide. International Journal of Ethics, 41(3):371–3.

Hind, A. M., 1999. Co-operative life cycle and goals. *Journal of Agricultural Economics*, 50(3):536–48.

Hingley, M., Mikkola, M., Canavari, M. and Asioli, D., 2011. Local and sustainable food supply: The role of European retail consumer co-operatives. *International Journal on Food System Dynamics*, 2(4):340–56.

Hinrichs, C., 2000. Embeddedness and local food systems: Notes on two types of direct agricultural market. *Journal of Rural Studies*, 16(3):295–303.

Hirschman, A.O., 1970. *Exit, Voice and Loyalty: Responses to decline in firms, organizations and states*. Harvard: Harvard University Press.

Hobsbawm, E. J., 1999. *Industry and Empire: From 1750 to the present day*. New York: The New Press.

Holloway, L., Kneafsey, M., Venn, L. et al., 2007. Possible food economies: A methodological framework for exploring food production–consumption relationships. *Sociologia Ruralis*, 47(1):1–19.

Holloway, L., Cox, R., Kneafsey, M. et al., 2010. Are you alternative? "Alternative" food networks and consumers' definitions of alterity. In Fuller, D., Jonas, A. E. and Lee, R. (eds), 2010. *Interrogating Alterity: Alternative economic and political spaces*. Farnham, UK: Ashgate.

Holt-Giménez, E. and Burkett, B., 2011. Food movements unite. In *Strategies to Transform Our Food Systems*. New York City: Food First Books.

Home, R., 2009. Land ownership in the United Kingdom: Trends, preferences and future challenges. *Land Use Policy*, 26:103–8.

House of Lords EU Committee, 2014. *Counting the cost of food waste: EU food waste prevention*. 10th Report of Session 2013–14. London: The Stationery Office

Huggins, C. D., 2014. Control grabbing and small-scale agricultural intensification: Emerging patterns of state-facilitated "agricultural investment" in Rwanda. *Journal of Peasant Studies*, 41(3):365–84.

ICA, 1995. *Agricultural co-ops in North America and Europe (UN/ECE 1995)*. International Cooperative Alliance. Available: www.uwcc.wisc.edu/icic/today/ag/na-europe.html

ICA, 2011. *Global 300 Report 2010*. International Cooperative Alliance. Available: http://ica.coop/sites/default/files/media_items/Global300Report2011.pdf

ICA, 2012. *Friedrich Wilhelm Raiffeisen*. International Cooperative Alliance. Available: http://ica.coop/en/history-co-op-movement/friedrich-wilhelm-raiffeisen

ICA, 2015. *Co-operative identity, values and principles*. International Cooperative Alliance. Available: http://ica.coop/en/whats-co-op/co-operative-identity-values-principles

ILO Co-operative Branch, 2012. Sustainable energy co-operatives (draft). Geneva: ILO. In: ICA, 2013. *Blueprint for a co-operative decade*. International Cooperative Alliance. Available: https://ica.coop/sites/default/files/media_items/ICA%20Blueprint %20-%20Final%20version%20issued%207%20Feb%2013.pdf

Iliopoulos, C., 2005. New generation cooperatives: The potential of an innovative institutional arrangement for Mediterranean food supply chains. *New Medit*, 4(1):14–20.

Illich, I., 1976. *Deschooling Society*. London: Pelican.

Innovate UK., 2015. *New agrimetrics centre will boost food and farming industries*. Available: www.gov.uk/government/news/new-agrimetrics-centre-will-boost-food-and-farming-industries

IPCC, 2014. *Climate change 2014: Impacts, adaptation, and vulnerability*. Intergovernmental Panel on Climate Change. Available: www.ipcc.ch/report/ar5/wg2/

Jaffee, D., 2010. Fair trade standards, corporate participation, and social movement responses in the United States. *Journal of Business Ethics*, 92(2):267–85.

Janzen, T., 2015. Does your co-op own your farm data? Available: http://www.linkedin.com/pulse/does-your-co-op-own-data-todd-janzen/

Jochnowitz, E., 2001. Edible activism: Food, commerce, and the moral order at the Park Slope Food Coop. *Gastronomica*, 1(4):56–63.

Johnson, C., 2014. Michel Bauwens on the rise of multi-stakeholder cooperatives. *Shareable*. Available: www.shareable.net/blog/michel-bauwens-on-the-rise-of-multi-stakeholder-cooperatives

Jones, O. et al., 2010. The alternativeness of alternative food networks: Sustainability and the co-production of social and ecological wealth. In Fuller, D., Jonas, A. and Lee, R. (eds) *Interrogating Alterity: Alternative economic and political spaces*, pp. 95–109. Farnham, UK: Ashgate.

Jones, D. and Kalmi, P., 2012. Economies of scale versus participation: A co-operative dilemma? *Journal of Entrepreneurial & Organizational Diversity*, 1(1):37–64.

Jones, A. D., Hoey, L., Blesh, J., Miller, L., Green, A. and Shapiro, L. F., 2016. A systematic review of the measurement of sustainable diets. *Advances in Nutrition: An International Review Journal*, 7(4): 641–64.

Julia, J. F. and Melia, E., 2003. Challenges for agricultural co-operatives in the European Union: The case of the Spanish agricultural co-operatives. *International Journal of Co-operative Management*, 1:16–23.

Juliá Igual, J. F. and Marí Vidal, S., 2002. Farm cooperatives and the social economy: The case of Spain. *Journal of Rural Cooperation*, 30(2):119–33.

JWT Intelligence, 2014. *Uniqlo, H&M and retail as the third space*. Available: www.jwtintelligence.com/2014/04/uniqlo-hm-retail-space/#ixzz39jDymE4f

Kalmi, P., 2007. The disappearance of cooperatives from economics textbooks. *Cambridge Journal of Economics*, 31(4):625–47.

Karantininis, K. and Nilsson, J., 2007. *Vertical Markets and Cooperative Hierarchies: The role of cooperatives in the agri-food industry*. Dordrecht, Netherlands: Springer.

Kassam, A., 2014. Spain's "Robin Hood" swindled banks to help fight capitalism. *The Guardian*. Available: www.theguardian.com/world/2014/apr/20/spain-robin-hood-banks-capitalism-enric-duran

Katz, S. E., 2006. *The Revolution Will Not Be Microwaved: Inside America's underground food movements*. White River Junction, VT: Chelsea Green.

Kay, A., 2003. Path dependency and the CAP. *Journal of European Public Policy*, 10(3):405–20.

Khoury, C. K., Bjorkman, A. D., Dempewolf, H. et al., 2014. Increasing homogeneity in global food supplies and the implications for food security. *Proceedings of the National Academies of Science*, 111(1):4001–6.

Kindling Trust, 2012. Growing Manchester(s) Veg People – a guide to setting up a growers' and buyers' co-operative. Available: www.sustainweb.org/publications/?id=214

Kloppenburg, J., Lezberg, S. et al., 2000. Tasting food, tasting sustainability: Defining the attributes of an alternative food system with competent, ordinary people. *Human Organization*, 59(2):177–86.

Knapp, J. G., 1965. *An Analysis of Agricultural Co-operation in England*. London: Agric. Centre Co-op. Ass.

Kneafsey, M., 2015. Community-led culture economies of food. Conference presentation, *Community-Led Re-Design of Cultural Heritage*, 13 November, Hamburger Bahnhof, Museum für Gegenwart, Berlin.

Kneafsey, M., Cox, R., Holloway, L., Dowler, E., Venn, L. and Tuomainen, H., 2008. *Reconnecting Consumers, Producers and Food: Exploring alternatives*. Oxford: Berg.

Kneafsey, M., Owen, L., Bos, E., Broughton, K. and Lennartsson, M., 2016. Capacity building for food justice in England: The contribution of charity-led community food initiatives. *Local Environment*, 22(5):1–14.

Kneafsey, M., Venn, L., Schmutz, U., Balázs, B., Trenchard, L., Eyden-Wood, T., Bos, E., Sutton, G. and Blackett, M., 2013. *Short food supply chains and local food systems in the EU. A state of play of their socio-economic characteristics*. JRC Scientific and Policy Reports. Joint Research Centre Institute for Prospective Technological Studies, European Commission.

Kohler-Koch, B., 1999. The evolution and transformation of European governance. In Kohler-Koch, B. and Eising, R. (eds) *The Transformation of Governance in the European Union*, pp. 14–35. London: Routledge.

Koning, N., 1994. The Failure of Agrarian Capitalism: Agrarian politics in the UK, Germany, the Netherlands and the USA, 1846–1919. London and New York: Routledge.

KPMG, 2012. *Expect the unexpected: Building business value in a changing world*. KPMG. Available: www.kpmg.com/dutchcaribbean/en/Documents/KPMG%20Expect_the_Unexpected_ExctveSmmry_FINAL_WebAccessible.pdf

Kropotkin, P. P., 1902. *Mutual Aid* (1939 edn). Harmondsworth, UK: Penguin Books.

Kropotkin, P. P., 1906. *The Conquest of Bread*. Available: http://theanarchistlibrary.org/library/petr-kropotkin-the-conquest-of-bread

Kuhn, T. S., 1962. *The Structure of Scientific Revolutions* (2nd edn). Chicago: University of Chicago Press.

Kyriakopoulos, K., Meulenberg, M. and Nilsson, J., 2004. The impact of cooperative structure and firm culture on market orientation and performance. *Agribusiness*, 20(4):379–96.

Lamine, C., 2005. Settling shared uncertainties: Local partnerships between producers and consumers. *Sociologia Ruralis*, 45(4):324–45.

Lamine, C., Magda, D. and Amiot, M. J., 2017. Addressing ecological and health dimensions in agrifood systems transitions: An interdisciplinary and comparative perspective. XXVII European Society for Rural Sociology Congress On-line Proceedings, p. 261.

Land Workers' Alliance, 2015. *Farmstart project to boost London's food sovereignty.* Available: http://landworkersalliance.org.uk/2015/10/farmstart-project-to-boost-londons-food-sovereignty/

Lang, T., 2005. Food control or food democracy? Re-engaging nutrition with society and the environment. *Public Health Nutrition*, 8(6a):730–37.

Lang, T., 2007. Food security or food democracy? In *Pesticides News*. London: Pesticide Action Network.

Lang, T., 2010. From value-for-money to values-for-money? Ethical food and policy in Europe. *Environment & Planning A*, 42:1814–32.

Lang, T., Barling, D. and Caraher, M., 2009. *Food Policy: Integrating health, environment and society*. Oxford: Oxford University Press.

Lang, T. and Heasman, M. A., 2015. *Food Wars: The global battle for mouths, minds and markets*. London: Routledge.

LANTRA, 2011. *Agricultural factsheet 2010–2011*. Available: www.lantra.co.uk/Downloads/Research/Skills-assessment/Agriculture-v2-(2010-2011).aspx

Lawrence, G. and Dixon, J., 2015. The political economy of agrifood: Supermarkets. In Bonanno, A. and Busch, L. (eds) *Handbook of International Political Economy of Agriculture and Food*, pp. 213–31. Cheltenham, UK: Edward Elgar.

Lélé, S. M., 1991. Sustainable development: A critical review. *World Development*, 19(6):607–21.

Levidow, L., Pimbert, M. and Vanloqueren, G., 2014. Agroecology in Europe: Conforming or transforming the dominant agro-food regime. *Agroecology & Sustainable Food Systems*, 38(10):1127–55.

Leviten-Reid, C. and Fairbairn, B., 2011. Multi-stakeholder governance in cooperative organizations: Toward a new framework for research? *Canadian Journal of Nonprofit & Social Economy Research*, 2(2):25–36.

Lincoln, Y. S. and Guba, E. G., 1985. *Naturalistic Inquiry*, vol. 75. Beverly Hills, CA: Sage.

Lindsay, G. and Hems, L., 2004. The arrival of social enterprise within the French social economy. *Voluntas: International Journal of Voluntary & Nonprofit Organizations*, 15(3):265–86.

Little, R., Maye, D. and Ilbery, B., 2010. Collective purchase: Moving local and organic foods beyond the niche market. *Environment & Planning, A*, 42(8):1797–813.

Lockie, S., 2009. Responsibility and agency within alternative food networks: Assembling the "citizen consumer". *Agriculture & Human Values*, 26(3):193–201.

Lund, M., 2011. *Solidarity as a Business Model: A multi-stakeholder cooperatives' manual.* Kent, OH: Cooperative Development Center, Kent State University.

Lund, M., 2012. Multi-stakeholder co-operatives: Engines of innovation for building a healthier local food system and a healthier economy. *Journal of Co-operative Studies*, 45(1):32–45.

Lymbery P. and Oakeshott, I., 2014. *Farmageddon: The true cost of cheap meat.* London: Bloomsbury.

McMichael, A., Powles, J. et al., 2007. Food, livestock production, energy, climate change and health. *The Lancet*, 370(9594):1253–63.

McNeill, J. R., 2003. *The Mountains of the Mediterranean World.* Cambridge: Cambridge University Press.

MAGRAMA, 2012. Anteproyecto de Ley de fomento de la integración de cooperativas y de otras entidades asociativas de carácter agroalimentario. Madrid: Ministerio de Agricultura, Alimentación y Medio Ambiente.

MAGRAMA, 2013. Proyecto de ley de fomento de la integración de cooperativas y de otras entidades asociativas de carácter agrolimentario. Madrid: Ministerio de Agricultura, Alimentación y Medio Ambiente.

Majuelo Gil, E., 2001. El cooperativismo católico agrario durante el franquismo. El caso navarro (1939–1975). In Lopez Villaverede, A. and Ortiz Heras, M. (eds) *Entre surcos y arados. El asociacionismo agrario en la España del siglo 20th*, pp. 137–70. Cuenca, Spain: Universidad Castilla La Mancha.

Majuelo Gil, E. and Pascual, A., 1991. Del catolicismo agrario al cooperativismo empresarial. Setenta y cinco años de la Federación de Cooperativas Navarras (1910–1985). Madrid: Servicio de Publicaciones del MAPA.

Manchester Veg People, 2014. *Manchester Veg People is something different.* Available: http://vegpeople.org.uk/about

Manchester Veg People, 2016. *Manchester Veg People is something different.* Available: www.vegpeople.org.uk/associations/

Marsden, T., Banks, J. and Bristow, G., 2002. The social management of rural nature: Understanding agrarian-based rural development. *Environment & Planning A*, 34(5):809–25.

Marsden, T. and Sonnino, R. 2012. Human health and wellbeing and the sustainability of urban–regional food systems. *Current Opinion in Environmental Sustainability*, 4:427–30.

Marx, K. 1866. *Instructions for the delegates of the Provisional General Council – the different questions.* Available: www.marxists.org/archive/marx/works/1866/08/instructions.htm#05

Mason, P. and Lang, T., 2017. *Sustainable Diets: How ecological nutrition can transform consumption and the food system.* Abingdon, UK: Taylor & Francis.

Maté, V. and Carlón, M., 1984. Cooperativas dentro de un orden. *Agricultura Revista Agropecuaria*, 625:610–15.

Meliá-Martí, E. and Martínez, A. M., 2014. Caracterización y análisis del impacto y los resultados de las fusiones de cooperativas en el sector agroalimentario español. Almeria: Universidad Almería.

Mellanby, K., 1975. *Can Britain Feed Itself?* London: Merlin.

Ménard, C. and Klein, P. G., 2004. Organizational issues in the agrifood sector: Toward a comparative approach. *American Journal of Agricultural Economics*, 86(3):750–5.

Merrett, C. D. and Walzer, N. (eds), 2004. *Cooperatives and Local Development: Theory and applications for the 21st century*. New York: M.E. Sharpe.

Miles, M. B. and Huberman, A. M., 1994. *Qualitative Data Analysis: An expanded source book* (2nd edn). Thousand Oaks, CA: Sage

Milio, N., 1976. A framework for prevention: Changing health-damaging to health-generating life patterns. *American Journal of Public Health*, 66(5):435–9.

Miller, A. and Lipman, M., 2012. *The Convolutions of Historical Politics*. Budapest: Central European University Press.

Moje, E. B., Ciechanowski, K. Mc. et al., 2004. Working toward third space in content area literacy: An examination of everyday funds of knowledge and discourse. *Reading Research Quarterly*, 39(1):38–70.

Mole Valley Farmers, 2014. *Mole Valley Farmers, Annual report and accounts*. Available: www.molevalleyfarmers.com/cms-webapp/userfiles/file/Annual%20Accounts/Annual%20Accounts%202013_14.pdf

Mole Valley Farmers, 2016. *About us*. Available: www.molevalleyfarmers.com/mvf/info/general/About_Us

Mollison, B. and Slay, R. M., 1991. *Introduction to Permaculture*. Tyalgum, NSW: Tagari.

Moog, S, Spicer, A. and Böhm, S., 2015. The politics of multi-stakeholder initiatives: The crisis of the Forest Stewardship Council. *Journal of Business Ethics*, 128(3): 469–93.

Mooney, P. H., 2004. Democratizing rural economy: Institutional friction, sustainable struggle and the Cooperative Movement. *Rural Sociology*, 69(1):76–98.

Mooney, P. H. and Majka, T. J., 1995. *Farmers' and Farmworkers' Movements: Social protest in American agriculture*. Social movements past and present. New York: Twayne.

Mooney, P. H., Roahrig, J. and Gray, T., 1996. The de/repoliticization of cooperation and the discourse of conversion. *Rural Sociology*, 61(4):559–76.

Moragues-Faus, A. M. and Sonnino, R., 2012. Embedding quality in the agro-food system: The dynamics and implications of place-making strategies in the olive oil sector of Alto Palancia, Spain. *Sociologia Ruralis*, 52(2):215–34.

Morales Gutierrez, A. C., Romero Atela, T. and Munoz Duenas, M. D., 2005. Historical records of 20th century agricultural co-operative movement in Europe: A comparative synthesis in the European Union. In Chaves, R., Monzon, J. L., Stryjan, Y., Spear, R. and Karafolas, S. (eds), *The Future of Co-operatives in a Growing Europe*, pp. 719–35. Valencia, Spain: CIRIEC-España.

Morgan, E., Tallontire, A. and Foxon, T. J., 2015. Large UK retailers' initiatives to reduce consumers' emissions: A systematic assessment. *Journal of Cleaner Production*. Available: http://sro.sussex.ac.uk/57318/

Morgan, K. and Sonnino, R., 2010. The urban foodscape: World cities and the new food equation. *Cambridge Journal of Regions, Economy & Society*, 3(2):209–24.

Morley, J., 1975. *British Agricultural Cooperatives*. London: Hutchinson.

Moulds, J. and Treanor, J., 2014. Co-op sells farms business to Wellcome Trust. *The Guardian*. Available: www.theguardian.com/business/2014/aug/04/co-op-sells-farms-business-wellcome-trust

Moyano Estrada, E., 1988. Sindicalismo y política agraria en Europa. Las organizaciones profesionales agrarias en Francia, Italia y Portugal. Madrid: Ministerio de Agricultura, Pesca y Alimentación.

Moyano Estrada, E., 2001. Acción colectiva y sindicalismo en la agricultura. In Bernal, A., Heras, M. O., Villaverde, A. L. (eds) *Entre surcos y arados: El asociacionismo agrario en la España del siglo 20th*, pp. 99–136. Cuenca, Spain: Universidad de Castilla La Mancha.

Muller, M. J., 2007. Participatory design: The third space in HCI. In Sears, A. and Jacko, J. A. (eds) *Human–Computer Interaction: Development process*, pp. 165–85. Boca Raton, FL: CRC Press.

Mulqueen, T., 2011. When a business isn't a business: Law and the political in the history of the United Kingdom's co-operative movement. *Oñati Socio-Legal Series*, 2(2):36–56.

Mulqueen, T., 2012. When a business isn't a business: Law and the political in the history of the United Kingdom's co-operative movement. *Oñati Socio-Legal Series*, 2(2):36–56.

Münkner, H., 2004. Multi-stakeholder co-operatives and their legal framework. In Borzaga, C. and Spear, R. (eds) *Trends and Challenges for Co-operatives and Social Enterprises in Developed and Transition Countries*, pp. 49–82. Trento, Italy: Edizioni31.

Murray, G. C., 1983. Towards an agricultural co-operative classification. *Journal of Agricultural Economics*, 34(2):151–61.

Naubauer, Ş., 2013. Predecessors and perpetrators of cooperative systems in Europe. *Lex ET Scientia International Journal (LESIJ)*, 20(1):40–50.

NEF, 2011. *The Ratio: Common sense controls for executive pay and revitalising UK business*. London: New Economics Foundation.

Nilsson, J., 1999. Co-operative organisational models as reflections of the business environments, *LTA*, 4(99):449–70.

Nilsson, J., Svendsen, G. L. and Svendsen, G. T., 2012. Are large and complex agricultural cooperatives losing their social capital? *Agribusiness*, 28(2):187–204.

Nowak, M. and Highfield, R., 2011. *SuperCooperators: Altruism, evolution, and why we need each other to succeed*. New York: Free Press.

Oakeshott, R., 1978. *The Case for Workers' Co-ops*. London: Routledge & Kegan Paul.

O'Connell, J., 1980. Report drawn up on behalf of the Committee on the Environment, Public Health and Consumer Protection on the Communication from the Commission of the European Communities to the Council (Doc. 222/79). Working Documents 1980–1981, Document 1-450/80, 8 October 1980.

OrganicLea, 2015. *Farmstart*. Available: www.organiclea.org.uk/we-work-for-change/farm-start/

Ortmann, G. F. and King, R. P., 2007. Agricultural cooperatives I: History, theory and problems. *Agrekon*, 46(1):18–46.

OSCAE, 2014. *Macromagnitudes del Cooperativismo agroalimentario español 2013*. Available: www.agro-alimentarias.coop/5/5_3_2.php

OSCAE, 2015. *Macromagnitudes del Cooperativismo agroalimentario español 2014*. Available: www.agro-alimentarias.coop/5/5_3_2.php

Ostrom, E., 1990. Governing the Commons: The evolution of institutions for collective action. Cambridge: Cambridge University Press.

Ostrom, E., 2009. Collective action theory. In Boix, C. and Stokes, S. (eds) *The Oxford handbook of comparative politics*. Oxford: Oxford University Press.

Oustapassidis, K., 1988. Structural characteristics of agricultural cooperatives in Britain. *Journal of Agricultural Economics*, 39(2):231–42.

Oxfam, 2012. *Tipping the Balance: Policies to shape agricultural investments and markets in favour of small-scale farmers.* Oxford: Oxfam International.

Oxford Farming Conference, 2014. *Opportunity Agriculture, the Next Decade – Towards a sustainably competitive industry.* Available: www.ofc.org.uk/files/ofc/papers/ofcreportonline.pdf

Oxford Real Farming Conference, 2016. *About.* Available: http://orfc.org.uk/about/

P2P Foundation, 2016. *Peer 2 Peer Foundation.* Available: http://p2pfoundation.net

P2P Value, 2016. Peer Value: Advancing the Commons Collaborative Economy. Amsterdam, 2–3 September.

Paine, T., 2008. *Rights of Man, Common Sense, and Other Political Writings.* Oxford: Oxford University Press (originally published 1791).

Paniagua, X., 1982. La sociedad libertaria. Agrarismo e industrialización en el anarquismo español (1930–1939). Barcelona: Editorial Críticas.

Paquet, G. 2009. Quantophrenia. *Optimum Online,* 39(1):14–27.

Patel, R., 2007. *Stuffed and Starved: Markets, power and the hidden battle for the world food system.* Melbourne: Black Inc.

Pathak, S. D., Wu, Z. and Johnston, D., 2014. Toward a structural view of co-opetition in supply networks. *Journal of Operations Management,* 32(5):254–67.

Peeters, B., 2011. Permaculture as alternative agriculture. *Kasarinlan: Philippine Journal of Third World Studies,* 26(1):422–34.

Peña, D. G. and Foucault, M., 2005. *Farmers Feeding Families: Agroecology in South Central Los Angeles.* Berkeley: University of California–Berkeley.

Pérez Neira, D. and Soler Montiel, M., 2013. Agroecología y ecofeminismo para descolonizar y despatriarcalizar la alimentación globalizada. *Revista internacional de pensamiento político,* 8:95–113.

Pimbert, M. P., Thompson, J., Vorley, W. T., Fox, T., Kanji, N. and Tacoli, C., 2001. *Global Restructuring, Agri-Food Systems and Livelihoods.* London: International Institute for Environment and Development (IIED).

Pimentel, D., Hurd, L. E. et al., 1973. Food production and the energy crisis. *Science,* 182(4111):443–9.

Pirkey, M. F., 2015. People like me: Shared belief, false consensus, and the experience of community. *Qualitative Sociology,* 38(2):139–64.

Pirson, M. and Turnbull, S., 2011. Toward a more humanistic governance model: Network governance structures. *Journal of Business Ethics,* 99(1):101–14.

Planas, J. and Valls-Junyent, F., 2011. ¿Por qué fracasaban las cooperativas agrícolas? Una respuesta a partir del análisis de un núcleo de la Cataluña rabassaire. *Investigaciones de Historia Económica,* 7(2):310–21.

Planas Maresma, J., 2008. Los propietarios rurales y la organización de la acción colectiva agraria al inicio del siglo 20th. Congreso de Historia Agraria, 13–15 May, Cordoba, Spain.

Planells Orti, J. and Mir Piqueras, J., 2004. Grupo empresarial Anecoop, Origen y Desarrollo. In Juliá Igual, J. F. (ed.) *Colección Mediterráneo Económico: Economía Social. La actividad económica al servicio de las personas* 6:386–403.

Plunkett Foundation, 1953. *Yearbook of Agricultural Cooperation 1953.* Oxford: Plunkett Foundation.

Plunkett Foundation, 1954. *Yearbook of Agricultural Cooperation 1954.* Oxford: Plunkett Foundation.

Plunkett Foundation, 1961. *Yearbook of Agricultural Cooperation 1966.* Oxford: Plunkett Foundation.

Plunkett Foundation, 1966. *Yearbook of Agricultural Cooperation 1966*. Oxford: Plunkett Foundation.

Plunkett Foundation, 1967. *Yearbook of Agricultural Cooperation 1967*. Oxford: Plunkett Foundation.

Plunkett Foundation, 1985. *Yearbook of Agricultural Cooperation 1985*. Oxford: Plunkett Foundation.

Polleta, F., 2002 *Freedom Is an Endless Meeting: Democracy in American social movements*. Chicago: University of Chicago Press.

Ponterotto, J. G., 2005. Qualitative research in counseling psychology: A primer on research paradigms and philosophy of science. *Journal of Counseling*, 52(2):126–36

Popkin, B., 2003. The nutrition transition in the developing world. *Development Policy Review*, 21(5–6):581–97.

Pratt, J. and Luetchford, P., 2014. *Food for Change: The politics and values of social movements*. London: Pluto Press.

Pretty, J., 2001. *Some benefits and drawbacks of local food systems*. Briefing Note for TVU/ Sustain AgriFood Network. Available: www.sustainweb.org/pdf/afn_m1_p2.pdf

Prothero, R. E., Ernle, B. et al., 1961. *English Farming, Past and Present*. London: Heinemann.

Rayner, A. and Ennew, C. T., 1987. Agricultural co-operation in the UK: A historical review. *Agricultural Administration & Extension*, 27(2)93–108.

Rayner, G. and Lang, T., 2012. *Ecological Public Health: Reshaping the conditions for good health*. Abingdon, UK: Earthscan/Routledge.

REAS, 2011. *Carta de Principios de la Economia Solidaria. Red de Redes de Economia Alternative y Solidaria*. Available: www.economiasolidaria.org/carta.php

Red de Cooperativas Integrales, 2016. *Ecoredes*. Available: http://integrajkooperativoj. net/somos/ecoredes/

Renting, H., Marsden, T. K. and Banks, J., 2003. Understanding alternative food networks: Exploring the role of short food supply chains in rural development. *Environment & planning A*, 35(3):393–411.

ResPublica, 2012. *Supplementary written evidence from ResPublica*. Available: www. publications.parliament.uk/pa/cm201213/cmselect/cmcomloc/writev/112/112.pdf

Reymond, P., 1964. Integration and consumers' cooperative organisations. *World Agr.*, 13(4):27–9.

Richardson, L., 2000. Writing: A method of inquiry. In Denzin, N. K. and Lincoln. Y. S. (eds) *Handbook of Qualitative Research*, pp. 923–48. Thousand Oaks, CA: Sage.

Rhodes, R., 2012. *Empire and Co-operation: How the British Empire used co-operatives in its development strategies, 1900–1970*. Edinburgh: John Donald.

Rockström, J., Steffen, W., Noone, K. et al., 2009a. Planetary boundaries: Exploring the safe operating space for humanity. *Ecology & Society*, 14(2):32.

Rockström, J., Steffen, W., Noone, K. et al., 2009b. A safe operating space for humanity. *Nature*, 461(7263):472–5.

Ronco, W., 1974. *Food Co-ops: An alternative to shopping in supermarkets*. Boston, MA: Beacon Press.

Rowe, F., 2011. Towards a greater diversity in writing styles, argumentative strategies and genre of manuscripts. *European Journal of Information Systems*, 20(5):491–5.

Sacks, J., 2013. *Sticky Money: Evaluating the local impact of the co-operative pound*. Manchester: Co-operatives UK. Available: www.uk.coop/sites/storage/public/ downloads/lm3_new_insight_10.pdf

Sanchez Bajo, C. and Roelants, B., 2011. *Capital and the Debt Trap: Learning from co-operatives in the global crisis*. New York: Palgrave Macmillan.

Sánchez Escolano, L. M., 2013. Modelo territorial innovador y articulación urbana en el poniente almeriense. *Investigaciones Geográficas*, 59:57–74.

Saravia Ramos, P., 2011. Las cooperativas agroecológicas como una alternativa a la producción, distribución y consumo de alimentos. *Papeles de relaciones ecosociales y cambio global*, 115:149–58.

Schonhardt-Bailey, C. (ed.), 1996. *Free Trade: The repeal of the Corn Laws*, vol. 10. Bristol: Thoemmes Press.

SAOS, 2012. *Annual Report 2011*. Edinburgh: Scottish Agricultural Organisation Society.

SAOS, 2017. *Annual Report 2017*. Edinburgh: Scottish Agricultural Organisation Society.

Serrano, R. et al., 2015. The internationalisation of the Spanish food industry: The home market effect and European market integration. *Spanish Journal of Agricultural Research*, 13(3):1–4.

Schindler, S., 2012. Of backyard chickens and front yard gardens: The conflict between local governments and locavores. *Tulane Law Review*, 87(2):231–96.

Schmitt, B. H., 2003. *Customer Experience Management: A revolutionary approach to connecting with your customers*. New Jersey: John Wiley.

Schneider A., Friedl, M.A., Potere, D. 2010. Mapping global urban areas using MODIS 500-m data: new methods and datasets. *Remote Sens. Environ.*, 114, 1733–1746.

Sennett, R., 2012. *Together: The rituals, pleasures and politics of cooperation*. London: Penguin.

Shaffer, J., 1999. *Historical Dictionary of the Cooperative Movement*. London: Scarecrow Press.

Sharzer, G., 2012. *No Local: Why small-scale alternatives won't change the world*. Winchester: Zero.

Seyfang, G., 2006. Ecological citizenship and sustainable consumption: Examining local organic food networks. *Journal of Rural Studies*, 22(4):383–95.

Simpson, J., 2003. *Spanish Agriculture: The long siesta, 1765–1965*. Cambridge: Cambridge University Press.

Simpson, J., 2005. Spanish agriculture in the long run, 1760–1960. In Jerneck, M., Mörner, M., Tortella, G. and Åkerman, S. (eds) *Different Paths to Modernity: A Nordic and Spanish perspective*, p. 74. Lund, Sweden: Nordic Academic Press.

Skogstad, G., 1998. Ideas, paradigms and institutions: Agricultural exceptionalism in the European Union and the United States. *Governance*, 11(4):463.

Slaughter, S. and Larry L., 1997. *Academic Capitalism: Politics, policies, and the entrepreneurial university*. Baltimore: Johns Hopkins University Press.

Smith, J., Machell, G. and Caraher, M., 2012. *MLFW Food Co-ops evaluation: Final report*. Available: www.city.ac.uk/__data/assets/pdf_file/0008/167138/MLFW-final-report-Food-Co-ops-Evaluation-2012-Feb-2013.pdf

Sober, E. and Wilson, D. S., 1998. *Unto Others: The evolution and psychology of unselfish behaviour*. Cambridge, MA: Harvard University Press.

Soberania Alimentaria, 2013. Reflexiones sobre el Cooperativismo Agrario. Soberania Alimentaria, Biodiversidad y Culturas, no. 15.

Social Landscapes, 2018. *Permaculture design course – creating resilient urban communities*. Available: www.sociallandscapes.co.uk/events/permaculture-design-course-pdc2018-south-london

Soja, E. W., 1999. In different spaces: The cultural turn in urban and regional political economy. *European Planning Studies*, 7(1):65–75.

Somerset Cooperative Services, 2009. *The Somerset Rules – Register an IPS multi-stakeholder co-op.* Available: www.somerset.coop/somersetrules

Somerville, P., 2007. Co-operative identity. *Journal of Co-operative Studies*, 40(1):5–17.

Sonnino, R., 2009. Feeding the city: Towards a new research and planning agenda. *International Planning Studies*, 14(4):425–35.

Sorokin, P. A., 1956. *Fads and Foibles in Modern Sociology and Related Sciences.* Chicago: Henry Regnery. Cited in Paquet, G., 2009. Quantophrenia. *Optimum Online*, 39(1):14–27.

Spear, R., 2010. Religion and value-driven social entrepreneurship. In K. Hockerts, K., Mair, J. and Robinson, J. (eds) *Values and Opportunities in Social Entrepreneurship.* Houndmills, UK: Palgrave.

Spear, R., Westall, A. and Burnage, A., 2012. *Support for Farmers' Cooperatives; Country report: The United Kingdom.* Wageningen: Wageningen UR. Available: http://edepot. wur.nl/244819

SpinWatch, 2005. *The "Big Food" takeover of British agriculture.* Available: http://spinwatch. org/spaw/images/artwork/Big-food.pdf

Steffen, W., Richardson, K., Rockström, J., Cornell, S. E., Fetzer, I., Bennett, E. M., Biggs, R., Carpenter, S. R., de Vries, W., de Wit, C. A. and Folke, C., 2015. Planetary boundaries: Guiding human development on a changing planet. *Science*, 347(6223):1259855.

Steinfeld, H., Gerber, P., Wassenaar, T. et al., 2006. *Livestock's Long Shadow: Environmental issues and options.* Rome: Food and Agriculture Organisation.

Stern N., 2006. The Stern Review of the Economics of Climate Change. Final report. London: HM Treasury.

STEPS, 2015. Pathways approach. Available: http://steps-centre.org/methods/ pathways-approach/

STEPS, 2011. *The Pathways Approach of the STEPS Centre.* Available: http://steps- centre.org/wp-content/uploads/STEPS_Pathways_online1.pdf

Stock, P. V., et al., 2014. Neoliberal natures on the farm: Farmer autonomy and coop- eration in comparative perspective. *Journal of Rural Studies*, 36:411–22.

Stofferahn, C. W., 2010. South Dakota soybean processors: The discourse of con- version from cooperative to limited liability corporation. *Journal of Cooperatives*, 24:13–43.

Sullivan, S., Spicer, A. and Böhm, S., 2011. Becoming global (un)civil society: Counter- hegemonic struggle and the indymedia network. *Globalizations*, 8(5):703–17.

Sustainable Development Commission, 2011. *Looking Back, Looking Forward: Sustainability and UK food policy 2000–2011.* London: Sustainable Development Commission.

Sustainable Food Trust, 2017. *The hidden cost of UK food.* Available: http://sustainablefood trust.org/articles/hidden-cost-uk-food/

Sutherland, N., Land, C. and Böhm, S., 2014. Anti-leaders(hip) in social move- ment organizations: The case of autonomous grassroots groups. *Organization*, 21(6):759–81.

Szabó, G. G., 2006. Co-operative identity, a concept for economic analysis and evalu- ation of co-operative flexibility: The Dutch practice and the Hungarian reality in the dairy sector. *Journal of Co-operative Studies*, 39(3):11–26.

Tapia, F. J. B., 2012. Commons, social capital, and the emergence of agricultural cooperatives in early twentieth century Spain. *European Review of Economic History*, 16(4):511–28.

Taylor, P. L., 2005. In the market but not of it: Fair trade coffee and forest stewardship council certification as market-based social change. *World Development*, 33(1):129–47.

The Land, 2012, Can Britain farm itself? *The Land*. Available: www.thelandmagazine.org.uk/articles/can-britain-farm-itself-2

Third Space Café, 2012. *The Third Space Story*. Available: www.thirdspace.ie/story/

Thompson, E. P., 1991. *The Making of the English Working Class*, rev. edn. London: Penguin.

Thompson, J., 2014. *Retail as the Third Space*. JWT. Available: http://jwt.co.uk/news/retail-as-the-third-space.html

Thompson, J., Millstone, E., Scoones, I., Ely, A., Marshall, F., Shah, E. and Wilkinson, J., 2007. *Agri-food system dynamics: Pathways to sustainability in an era of uncertainty*. SETPS Working Paper 4. Brighton: University of Sussex.

Thrupp, L. A., 2000. Linking agricultural biodiversity and food security: The valuable role of agrobiodiversity for sustainable agriculture. *International Affairs*, 76(2):283–97.

Torgerson, R. C., Reynolds, B. J. and Gray, T. W., 1997. Evolution of cooperative thought, theory and purpose. *Cooperatives: Their Importance in the Future of the Food and Agricultural System Conference*. 16–17 January, Las Vegas, NV: Food and Agricultural Marketing Consortium.

Tornaghi, C., 2014. *How to Set Up Your Own Urban Agricultural Project with a Socioenvironmental Justice Perspective: A guide for citizens, community groups and third sector organisations*. Leeds: The University of Leeds

Tortia, E. C., Valentinov, V. and Iliopoulos, C., 2013. Agricultural cooperatives. *Journal of Entrepreneurial & Organizational Diversity*, 2(1):23–36.

Tracy, S. J., 2010. Qualitative quality: Eight "big-tent" criteria for excellent qualitative research. *Qualitative Inquiry*, 16(10):837–51.

Tudge, C., 2011. *Good Food for Everyone Forever: A people's takeover of the world's food supply*. Pari, Italy: Pari.

Tukker, A., Bausch-Goldbohm, S., Verheijden, M. et al., 2009. *Environmental Impacts of Diet Changes in the EU*. Seville: European Commission Joint Research Centre Institute for Prospective Technological Studies.

Tukker, A., Huppes, G., Guinée, J., et al., 2006. Environmental Impact of Products (EIPRO): Analysis of the life cycle environmental impacts related to the final consumption of the EU-25. Brussels: European Commission Joint Research Centre.

Turnbull, S., 2012. Discovering the "natural laws" of governance. *The Corporate Board*, March/April:1–5.

Turnbull, S., 2013. Adopting the laws of nature to protect nature? *Sustainable Companies Conference*. 5–6 December, Oslo.

UK Parliament, 2014. Co-operative and Community Benefit Societies Act 2014. Available: www.legislation.gov.uk/ukpga/2014/14/pdfs/ukpga_20140014_en.pdf

UN, 2011. World Economic and Social Survey 2011: The great green technological transformation. New York: United Nations Department of Economic and Social Affairs.

UNDP, 1996. *Human Development Report 1996: Economic growth and human development*. New York: Oxford University Press.

UNEP, 2009. The Environmental Food Crisis: The environment's role in averting future food crises. A UNEP rapid response assessment. Arendal, Norway: United Nations Environment Programme/GRID-Arendal.

USDA, 1997. *Strengthening ethics within agricultural cooperatives.* USDA Rural Business Cooperative Service, Research Report 151.

USDA, 2002. *Agricultural cooperatives in the 21st century.* Available: www.rurdev.usda. gov/rbs/pub/cir-60.pdf

USDA, 2014. Comparing Cooperative Principles of the U.S. Department of Agriculture and the International Cooperative Alliance. Washington, DC: United States Department of Agriculture.

van Bekkum, O., 2001. *Cooperative Models and Farm Policy Reform: Exploring patterns in structure–strategy matches of dairy cooperatives in regulated vs. liberalized markets.* Assen, Netherlands: Van Gorcum.

van Bekkum, O. and van Dijk, G., 1997. *Agricultural Co-operatives in the European Union: Trends and issues on the eve of the 21st century.* Assen, Netherlands: van Gorcum.

van der Ploeg, J. D., 2008. *The New Peasantries: Struggles for autonomy and sustainability in an era of empire and globalization.* London: Earthscan.

van der Ploeg, J. D., 2013. *Peasants and the Art of Farming: A Chayanovian manifesto.* Winnipeg: Fernwood.

Vanloqueren, G. and Baret, P., 2009. How agricultural research systems shape a technological regime that develops genetic engineering but locks out agroecological innovations. *Research Policy,* 38(6):971–83.

Vara Sanchez, I., 2008. Aprendiendo a elegir colectivamente: Cooperativas agroecológicas de producción, distribución y consume. *Ciclos,* 20:20–22.

Vasilachis de Gialdino, I., 2009. Ontological and epistemological foundations of qualitative research. *Forum: Qualitative Social Research,* 10(2), art. 30.

Via Campesina, 2011. *The international peasant's voice.* Available: https://viacampesina.org/ en/index.php/organisation-mainmenu-44/what-is-la-via-campesina-mainmenu-45

Vieta, M., 2010. The new cooperativism. *Affinities: A Journal of Radical Theory, Culture, & Action,* 4(1):1–11.

von Mises, 1990. *Observations on the cooperative movement. Mises Institute.* Available: https://mises.org/library/money-method-and-market-process/html/p/393

Voorhis, J., 1961. *American Cooperatives: Where they come from, what they do, where they are going.* New York: Harper.

Vorley, B., 2003. Food Inc. Corporate Concentration from Farm to Consumer. London: UK Food Group.

Wageningen, 2013. *Support for farmers' cooperatives.* Available: www.wageningenur.nl/ en/show/Support-for-Farmers-Cooperatives.htm

Walley, K. and Custance, P., 2010. Co-opetition: Insights from the agri-food supply chain. *Journal on Chain & Network Science,* 10(3):185–92.

Thompson, J., 2014. *Retail as the Third Space.* JWT. Available: http://jwt.co.uk/news/ retail-as-the-third-space.html

Wales Rural Observatory, 2011. *Farmers' decision making.* Available: www.walesrural observatory.org.uk/sites/default/files/Farmers'%20Decision%20Making%20 Final%20Report.pdf

WAOS, 2011. *Presentation WF Cross Party Group.* Available: www.cooperatives-wales.coop/wp-content/uploads/2011/10/waos-presentation-WG-Cross-Party-Group-2011.pptx

WCED, 1987. *Our Common Future.* World Commission on Environment and Development. London: Oxford University Press.

Webb, S. and Webb, B., 1897. *Industrial Democracy.* London: Longmans, Green.

Webb, S. and Webb, B., 1922. *The Consumers' Co-operative Movement*. Available: https://archive.org/stream/consumerscoopera00webbuoft/consumerscoopera00webbuoft_djvu.txt

Welsh Co-operative and Mutuals Commission, 2014. *Report of the Welsh Co-operative and Mutuals Commission*. Available: http://gov.wales/docs/det/publications/140221coopreporten.pdf

WHO, 2011. *Global status report on noncommunicable diseases 2010*. Rome: World Health Organization.

WHO, 2013. *Obesity and overweight: Factsheet 311*. Geneva: World Health Organization.

Whyman, P. B., 2012. Co-operative principles and the evolution of the "dismal science": The historical interaction between co-operative and mainstream economics. *Business History*, 54(6):833–54.

Wilson, G. A., 2008. From 'weak' to 'strong' multifunctionality: Conceptualising farm-level multifunctional transitional pathways. *Journal of Rural Studies*, 24(3):367–83.

Wilson, F. and MacLean, D., 2012. The Big Society, values and co-operation. *Work, Employment & Society*, 26(3):531–41.

WWF, 2013. *Thirsty Crops: Our food and clothes: eating up nature and wearing out the environment?* Zeist, Netherlands: WWF.

Wynne-Jones, S., 2017. Understanding farmer co-operation: Exploring practices of social relatedness and emergent affects. *Journal of Rural Studies*, 53:259–68.

Young, A. R., 1995. Participation and policy making in the European Community: Mediating contending interests. *ECSA 4th Biennial International Conference*, 1995, Charleston, SC: USA.

Zeuli, K., 2004. The evolution of the cooperative model. In Merrett, C. D. and Walzer, N. (eds) *Cooperatives and Local Development: Theory and applications for the 21st century*, pp. 52–69. New York: M.E. Sharpe.

Zeuli, K. and Cropp, R., 2004. *Cooperatives: Principles and practices in the 21st century*. Madison: University of Wisconsin. Available: http://community-wealth.org/sites/clone.community-wealth.org/files/downloads/report-zeuli.pdf

Zeuli, K. and Radel, J., 2005. Cooperatives as a community development strategy: Linking theory and practice. *Journal of Regional Analysis & Policy*, 35(1):43–54.

Index

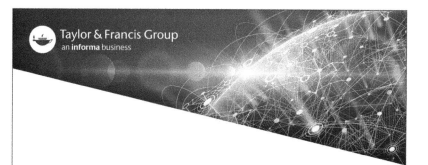

Milton Keynes UK
Ingram Content Group UK Ltd.
UKHW040109071024
449327UK00019B/924